MEAN GENES

MEAN GENES

From sex to money to food **Taming our primal instincts**

Terry Burnham and **Jay Phelan**

Perseus Publishing Cambridge, Massachusetts

Library of Congress Catalog Carad Number: 00-105183
ISBN 0-7382-0230-4

Perseus Publishing is a member of the Perseus Books Group.

Find us on the World Wide Web at http://www.perseuspublishing.com

Perseus Publishing books are available at special discounts for bulk purchases in the U.S. by corporations, institutions, and other organizations. For more information, please contact the Special Markets Department at HarperCollins Publishers, 10 East 53rd Street, New York, NY 10022, or call 1-212-207-7528.

Set in 10.5 pt Minion

First printing, August 2000

1 2 3 4 5 6 7 8 9 10 — 03 02 01 00

Contents

Contents

MEAN GENES

Introduction Our toughest battles are with ourselves

Consider this book an owner's manual for your brain.

Most of us would feel cheated if we bought a car or a microwave and it came without instructions. But our most important possessions — body and mind — come with no such guide, leaving us searching haphazardly for satisfaction: a dollop of exercise, thirteen minutes of sex, a "Happy Meal," a cocktail, and a sport-utility vehicle. *Mean Genes* offers the missing information that can help us take better control of our lives.

When we drive a car or operate a microwave, our orders are carried out exactly as we command. The machine doesn't talk back or have an agenda of its own — at least not yet. On the other hand, if we tell our brain, as part of a New Year's resolution, to cut down on fatty foods, it most likely will let out a hearty laugh and continue to set off bells and whistles of approval when the dessert cart rolls around.

Our brain, for better or worse, is not an obedient servant. It has a mind of its own. Imagine that you are actually two things:

a personality who has likes, dislikes, desires, and dreams. But inside your body there is also a "machine," your brain, that processes commands and acts on those likes, dislikes, desires, and dreams. It fights you all the time. And it usually wins.

Why can't the two of you see eye to eye? *Why do we have battles over controlling our own behavior? And why are these battles so hard to win?* Are cats and dogs obsessed with fighting addictions, controlling their weight, and remaining faithful to their mates? Do chimpanzees regularly resolve to be less selfish?

In a creepy campfire legend, a babysitter alone in a house receives increasingly menacing phone calls. Terrified, she contacts the police, who put a tap on her phone. After the boogieman calls again, the cops frantically phone her, screaming, "We've traced the call. It's coming from inside the house! Get out!"

Similarly, the source of our self-control problems lies within us, in our genes. But we can't get out or leave them behind. Manipulative media, greedy businesses, and even our friends and family play roles in nurturing our demons. Still, most of our self-control problems stem from our impulses to do things that are bad for us or for those whom we love.

A visit to any bookstore reveals the nature of our struggles. Glancing at the bestsellers, we can see what's on people's minds. There are dozens of books on finding love, losing weight, and creating wealth. Conspicuously absent are a host of other topics. Where are the books entitled *How to Build a Bigger Beer Gut, Ten Steps to Frivolous Spending,* or *Nurturing the Infidel Within*? Why do some behaviors come so naturally while others require so much effort? It's because our genes predispose us to certain failings.

Every day we see headlines heralding discoveries of "the gene for alcoholism" or "the genes of aging." These reports make an obvious point: human biology and disease are influenced in important ways by our genes. The Human Genome Project, which will soon determine the sequence of every stretch of DNA in the human body, is a revolution in the making. With each passing week, scientists unearth the genetic roots of more and more diseases, and the promise of future cures grows stronger.

But genetic effects are far more pervasive than these articles suggest. Even in areas where we feel that we act purely of our own free will, our dramas are played out on a genetic stage. Over the last few decades, scientists have learned a great deal about the structure of this stage, and our learning will accelerate with the forthcoming genomic advances. Throughout *Mean Genes,* we explore what we know about these genetic influences and what they mean for our daily lives. Let us entice you with one example:

What is beauty and who sets the standards? It's a complex question, and many before us have tried to answer it. Some have suggested that beauty is mysterious or divine, incapable of comprehension by mere mortals.

Others believe that beauty is defined by the society we live in: whatever the fashion industry deems attractive is considered beautiful by the public at large. But if this were true — if beauty were truly governed by fads or trends — wouldn't every culture have its own definition of beauty? This is not the case.

Careful observation of human symmetry shows us why. The two halves of our body mirror each other. Our right hand, for example, is structured exactly like our left hand. The mirroring is not perfect, though, and each of us is aware of our minor deviations from complete symmetry — one ear may be a bit lower, one breast slightly larger, etc.

We find symmetrical people beautiful, even if they are not "classically" attractive. In scientific studies, both women and men show a clear and dramatic preference for symmetrical partners over more lopsided lovers. Furthermore, women who recorded details of their sex lives revealed an interesting pattern: they were much more likely to reach orgasm and more likely to become pregnant during intercourse with symmetrical men.

The symmetry data are more than just a tantalizing tale: across the animal kingdom, symmetry is a sign of healthy, disease-free bodies, likely to have been built by a good set of genes. Although most of us cannot assess someone's symmetry, it guides our mating decisions unconsciously.

So there is a logic to our aesthetic penchant for symmetrical people — logic that only a gene could love, logic that can be understood only by looking at ourselves in the broader context of evolution and animal behavior. Doing so reveals that our brain *does* have an agenda, but this agenda need not remain a mystery.

Our brains have been designed by genetic evolution. Once we understand that design, it is no longer surprising that we experience tensions in our marriages, that our waistlines are bigger than we'd like, and that Big Macs are tastier than brown rice. To understand ourselves and our world, we need to look not to Sigmund Freud but rather to Charles Darwin.

Like it or not, we are each engaged in a battle against our own set of mean genes. They are wily opponents, too. Masters of the visceral, they control through satisfaction, pain, and pleasure.

Even the most successful people succumb. Look at Oprah Winfrey, for instance. She runs a powerful media empire and is reportedly closing in on billionaire status. Her long list of accomplishments includes seven Emmy awards, an Oscar nomination, and a beauty queen crown. Rich and influential, this exceptional person is also very ordinary in one respect. Along with the rest of us, Oprah struggles for self-control.

Because she has been so honest about her weight and other personal issues, Oprah has helped millions. Furthermore, because her journey has been so successful in spite of powerful urges, she demonstrates an important point: we are not lumbering robots doomed to carry out our genetic programming.

In daily life, two paths beckon. One tempts us simply to live as our urges and passions direct. This can be called the "pet path" since it is followed by all animals, including the family dog. Eat when hungry. Eat until the food is gone. Remain loyal and faithful only to the extent that loyalty and faithfulness pay. If something feels good, do it again. If something hurts, avoid it.

Less clearly marked is the alternative, the path of most resistance. On this path we take charge, calling our own shots. Along with passions, genes have created willpower and the ability to control behavior consciously. With these uniquely human abilities, we can rise above our animal instincts.

Mean Genes is a guide to doing just that. Step 1 is to understand our animal nature, particularly those desires that get us into trouble and can lead to unhappiness. Step 2 is to harness this knowledge so that we can tame our primal instincts.

As you read this book, you'll see that we, your authors Terry and Jay, take this all very personally. Over the years that we've researched and taught this material, *Mean Genes* has become much more than a book to us. Throughout, you'll find tales from our personal lives. *Mean Genes* is not some stuffy academic tome; understanding the theory and taking the practical steps we suggest can improve your life. We are sure it can help you because it has helped us and our friends.

We would all like to make progress instantly, but there are no shortcuts. In auto racing, for example, designers struggle to make lighter and lighter cars. Perhaps surprisingly, the best way to trim 100 pounds of weight from a car is to find 1,000 places to trim a tenth of a pound. No grand redesigns, no massive overhauls, just relentless striving for incremental improvement. Similarly, for most of us, the best way to improve our lives is to find numerous small ways to change for the better.

The *Mean Genes* approach is not going to solve all of our problems in a few days. Rather, we think of it as a pair of glasses, enabling us to see the world more clearly. Corrective lenses don't change the basic struggle. We both still want to be leaner and nicer and have more friends, for example. So while the world we live in remains constant, it makes more sense viewed through our *Mean Genes* spectacles.

This more accurate view of the world can help in concrete ways. Consider a recent conversation between Terry and one of his friends. Karen is a 32-year-old graduate student who, along with her husband, is thinking of starting a family. Karen wanted to lose a little weight before getting pregnant. Terry cautioned against this approach. When a woman wants to get pregnant, the best advice is actually to *gain* a few pounds. Why?

Our bodies are built to be sensitive to the environment. In particular, our distant ancestors lived in a world where food was so scarce that raising a baby was difficult, so it was important to become pregnant when times were relatively good. The solution, built into women's bodies, is that fertility is modulated by weight changes. Even minor weight loss caused by short-term dieting or exercise dramatically decreases fertility and can easily delay pregnancy for months.

This translates to practical advice. If a woman wants to get pregnant, she should eat normally and avoid losing weight. This is true for all women, regardless of their weight. In our quest for happier lives, this fertility tip is like one little improvement to our racecar.

Another incremental bit of progress comes from Jay's money-saving techniques. Each month, Jay completely drains his checking account. If the bank machine will give him money, he will spend it. It's not that Jay is particularly weak. He's human and shares our natural tendency to spend too much. Why?

Think back, once again, to that long period of our evolutionary history we spent as hunter-gatherers. We evolved in a world where wealth existed primarily in the form of food and could not be stored for very long — any surplus would rapidly spoil. So our brains were designed in an era when the best way to save was to consume. Is it any wonder that Jay's natural tendency is to deplete his surplus each month?

Through the *Mean Genes* specs, Jay has spied a solution. He has instructed his employer to withhold a big chunk of his pay each month. The deducted funds are still Jay's, but now he doesn't have easy access to them and therefore doesn't feel he's sitting on a surplus that will spoil. (The funds are squirreled away and cannot be squandered without at least a phone call and a few days' wait.) By hiding part of his paycheck from the overconsuming monster within, Jay lets his slightly inaccessible cash pile up for a rainy day.

~

Mean Genes seeks to foster a deep understanding of human existence, drawing from diverse disciplines and hundreds of sources. We garner insights, for example, by looking at a range of cultures, including many that are worlds apart from our own. We also learn about humans by studying animals ranging from our close genetic cousins, the chimpanzees, to mice and even fruit flies. But the foundation of the book is evolutionary biology.

Ever since Darwin published *The Origin of Species* in 1859, people have debated the role of biology in human affairs. Still, as the idea of evolution has itself evolved, one provocative facet has become clearer and clearer: the human brain has been shaped by evolution. From its tremendous size right down to the mechanisms by which individual neurons talk to each other, our brain — like our eyes, arms, legs, and kidneys — is a product of natural selection. We know this is true. Does it follow that our psyches, too, have been shaped by evolution?

We think so, but not everyone agrees. Some mock the idea; others are troubled or even angered by it. But volumes of research have begun to quiet the critics. The more progress that is made in unraveling the genes we carry, the more clear it becomes that our evolutionary inheritance plays a central role in our lives.

Are genes the whole story? Obviously not. Other factors are important in determining every human characteristic. We know, for example, that physical or emotional abuse can scar children, regardless of their genetic endowment. Similarly, while we each inherit a particular genetic risk for heart disease, lifestyle decisions affect our health dramatically.

In this book, we focus primarily on the genetic role. Many other books describe the cultural influences on the behaviors we address, and we encourage you to study the ways that genetic and environmental factors interact to shape our lives.

Being full-time academics, we read hundreds of obscure research articles each year. We attend conferences and eagerly line up for talks with such titles as "The Phylogeny of New World Monkeys" and "DNA Damage–Induced Activation of p53 by the Checkpoint Kinase Chk2." We debate with other scientists on the front lines, discussing breaking studies long before the information hits the *New York Times* and other media.

Most people don't spend their lives similarly immersed in scientific detail. But everyone can benefit from the knowledge about human nature that has incrementally — and relentlessly — piled up during the last forty years, a period known as the second Darwinian revolution. This knowledge has changed our lives. And we think it can help you live a rich and passionate life without self-destruction. Think of us as your translators, bringing the crucial information from the frontiers right into your living room.

We take our responsibilities as your translators of science very seriously. The chapters ahead are filled with many stories and data. Although they are free of scientific jargon, every fact has been assiduously researched. There are more than a thousand citations, verifying every aspect of the book. We haven't included them here — they would fill more pages than the text itself — but if you are curious or just want to learn more, a set of notes is available at www.meangenes.org.

Mean Genes is the first book that converts the modern Darwinian revolution into practical steps for better living. Some of our advice appears to be common sense. Often, though, viewing life through our *Mean Genes* glasses means taking unexpected and seemingly bizarre steps. Consider:

Why does Jay fill up on day-old, dry bagels as he heads to a gourmet meal at a friend's house? And what does he gain by quickly smearing mayonnaise on the brownie delivered with his lunch on every flight from Los Angeles to Boston?

The FedEx driver invariably rubs his head in confusion when Terry hands him a package for overnight delivery . . . back to Terry. It doesn't seem to help when Terry explains that the package contains the short cable that connects his computer to the Internet.

Despite his love of giving and receiving gifts, Jay claims that birthdays and holidays are absolutely the worst time to shower our loved ones with gifts. Amazingly, he still has friends and Lisa, his wife. Why?

We no longer try to shield our friends from our unorthodox code of behavior. We regularly urge them, for instance, to respect the Big Four. That is, to be unusually kind and attentive to their spouses for four days every month. Not just any four days, but a specific set of days more important than payday. Can you imagine which days they are? (Peek ahead to the discussion in "Romance and Reproduction" if you must.)

All of these behaviors come from seeing the world through the *Mean Genes* lenses that allow us to predict when we will be weak and why we are vulnerable. The twig of human nature is indeed bent from the start. It must be seduced, not bullied, into behaving. Battles for self-control are not defects of personality, nor can they be won in the sense that the foe is vanquished. To take control of our lives, we need perpetual vigilance and an understanding of the enemy within.

We invite you to read on and construct your own set of *Mean Genes* spectacles. Every person will have a slightly different prescription, but the overall aim is the same — to see our world more accurately so that we can control our instincts before they control us. In this way, we can lead more satisfying lives. Lives with integrity.

THIN WALLETS AND **FAT BODIES**

Debt Laughing all the way to the Darwinian bank

Why do we have such a hard time saving money? Take the following quiz: First, how much money would you like to save each month? Write down your answer as a percentage of your income. Second, how much money are you saving? Look at the last few months of your actual savings behavior, not your dreams about next year after you pay off your credit card debt. Write down your actual savings as a percentage of income. Now compare the two figures. The unpleasant reality is that most of us save far less than we want to.

Average Americans want to save 10% of their income and claim to save about 3%. If only that were true. We set a record low in February 2000, with a 0.8% savings rate. In other words, if you took home $2,000 after taxes and you saved like an average American, you spent every cent except a measly sixteen bucks.

The result is that Americans have little or no cash to spare. Enticed to spend by urgings everywhere we turn — from the Internet to billboards to crafty product placements on TV and in movies — we are a nation of spenders, rushing to deposit pay-checks into minuscule bank accounts to cover the checks we have written.

To understand our spending behavior, let's visit some of the world's most accomplished savers by taking a trip to northern Europe. There we find forests where autumn arrives much as it does throughout the temperate parts of the world. Leaves change their color, temperatures plummet, and winds pick up.

Look down as you walk through the forest and you'll see a feverish acknowledgment of the oncoming winter. Red squirrels shift into overdrive each September, forsaking their summer life of leisure. In the course of two months, each squirrel will hide more than three thousand acorns, pinecones, and beechnuts throughout the several acres of their home range. It's hard being a squirrel.

Come winter, however, diligence pays off. With little food to be found on the bare trees, some squirrels are still living large. Each day they methodically move from one storage spot to the next as they ultimately recover more than 80% of their stashed snacks, enough to keep them alive until spring.

Hoarding for the future isn't restricted to rodents with big cheeks. It's a common response throughout the animal kingdom when lean times are ahead. Many bird species also store food in the fall. Nutcrackers, for example, bury seeds from pine trees and, like squirrels, show remarkable memory in finding their savings.

If there were a Savings Hall of Fame, it would contain dozens of animal species but certainly not the average American. How can humans (at least most Americans) be so much worse at preparing for lean times than squirrels, birds, and an ark full of other dim-witted creatures?

As described in the fable of the grasshopper and the ant, there are two strategies for dealing with abundance. The grasshopper plays all summer long while the ant works relentlessly to store food. When winter comes, the ant survives and the grasshopper dies.

Similarly, squirrels that work hard to store nuts survive the winter to have babies in the spring. When those babies grow up, they have the genes of their parents, genes that tell them to start burying nuts when fall comes. Animals are accomplished savers because natural selection favors the appropriately thrifty. Shouldn't the same forces have produced frugal humans? To understand the answer, we can learn by observing the behavior of people who live as foragers, as our ancestors did until recently.

The !Kung San live in the deserts of southern Africa. Until the 1960s they lived off this harsh land as nomads, gathering plants and hunting animals much as their ancestors had for ten thousand years or more. Because some San were still hunting and gathering into the 1960s, we have detailed records of their behavior in circumstances similar to those of our ancestors.

The !Kung San perpetually faced uncertain supplies of water and food. Building up reserves for the future would certainly help buffer those risks. Did the !Kung San save? Absolutely. The best opportunity for this saving came in times of windfall, usually after the killing of a large animal like a giraffe. With hundreds of pounds of edible giraffe meat, a hunter with a good savings system could live for months.

But !Kung San hunters had no meat lockers or freezers. Even if they preserved the extra meat, neighbors would descend and devour even the largest kill in a few days. Imagine your own "popularity" if you won the lottery, and you've got a pretty good picture of a !Kung San hunter with a dead giraffe outside his hut.

The !Kung San's behavior provides the clue to resolving the paradox between Americans' chronic undersaving and the strong evolutionary pressure to prepare for lean times. In a world without refrigerators or banks, preparing for hard times means eating enough food to store some fat on your body.

Although many animals, including squirrels and birds, do store food in the environment, most animals save by storing fat. Consider the interesting species called the elephant seal. A fully grown male of this species, at thirteen feet long and over two tons, is frighteningly similar in size to a fully loaded Cadillac. Females weigh in at a more demure half ton.

Each year, as the mating season approaches, elephant seals bulk up, with males loading on as much as two thousand extra pounds of body fat. Then, in an act of stunning single-mindedness that makes spring break in Miami look like Bible camp, the seals head for shore, and for three entire months they forgo all food, looking instead for love.

How do they survive? They draw from their substantial savings account. Before the ordeal winds down, they will lose more than a third of their weight. A male may shed more than a ton of blubber (and father up to a hundred babies).

So elephant seals save up for the mating season by storing extra energy on their body as fat. Humans, unfortunately, also save in this way. If you are a man, look down at your waist and grab the flesh that covers your stomach. If you are a woman, look at your thighs and buttocks. What do you see?

From one perspective you see hated fat, but from an evolutionary perspective you are looking at a savings account with a substantial (and possibly growing) balance. Evolution has produced a world of accomplished savers; humans, like most animals, simply save in the currency of body fat.

How good an evolutionary saver are you? In 1981 Bobby Sands, a member of the Irish Republican Army, went on a hunger strike to protest British policy. He was not a fat man to begin with, yet it still took him sixty-six days to starve himself to death. Many of us would survive, albeit unpleasantly, for more than two months without a single morsel of food. That's a pretty impressive savings account! Perhaps we deserve a place in that Savings Hall of Fame after all.

Consider an ancestral human who had just won a prehistoric lottery. He or she has, for example, just killed a wild pig or found a tree bursting with juicy mongongo nuts (not too different from macadamia nuts). With today's markets and financial instruments, this winner could sell the surplus and put the resulting cash in a bank.

For our ancestors, however, saving through markets and money was not an option. Successful people would ram as much as possible into their own stomachs and those of their genetic relatives. They might also share with non-relatives who would repay them on their good days. In such an environment, the best way to save is, paradoxically, to consume. Rather than leave some precious energy lying around to mold or be stolen, put it in your stomach and have your body convert the food into an energy savings account.

When you're a mammal, food is the coin of the realm. Genetic mechanisms prod squirrels to mind their nuts and elephant seals to pad their flanks. As we struggle to save money, our mammalian heritage lurks in the background. We *know* we ought to put some money in the bank, but consuming just *feels* so good.

⌐

Don't eat the nest egg. Proud as we may be of our hardy forebears and their genetic legacy, most of us would be happier if we could act less like victorious cavewomen and more like Scrooge. To prosper in the industrialized world with its refrigerators and government-insured bank accounts, we need to trick our ancient genes.

Because we evolved to consume everything in sight, many of the most successful savings techniques involve hiding money from ourselves. By making ourselves feel poor, we can induce our overconsuming instincts into living more frugally. One well-known technique is carrying less cash. By doing this we fool our genes, at least a bit, into thinking there is less surplus to be consumed.

In the movie *The Border,* Jack Nicholson comes home to find his house filled with expensive new furniture. When he asks how much it cost, his wife replies, "We don't have to worry 'bout payin' . . . I opened up a charge account!" A danger of using credit is that we do not hand over anything that feels valuable (such as cold, hard cash), so charging doesn't feel like spending money. In the quest for restraint, a credit card is worse than a debit card, a debit card worse than paying with cash, and paying with cash worse than not spending.

As a variation on this ruse, many people find multiple bank accounts useful. One account is untouchable and can accumulate savings, while the more commonly used account, usually the checking account, gets fixed transfers each month. The savings account should be as hidden as possible. For example, it can be in another state with no associated ATM or debit card. Or at least in a bank with ATMs located only in distant locations.

Easy access is our enemy. Ironic as it is, the best bank for our savings may be the one that makes withdrawals as difficult as possible. We can, for example, choose an account that pays a high interest rate but has outrageous fees for every transaction.

People don't come into the world with instincts for appropriate financial behavior. Most of us need to learn, and this learning frequently involves some painful mistakes. We (Terry and Jay) have been there, so we know.

Early in his financial life, Jay discovered the joy of credit cards. Freed from the tiresome need to have actual cash for purchases, he enjoyed an extended spending spree. But he soon learned that credit-fueled feasts end with maxed-out accounts and monster monthly payments that barely make dents in the balances. Each purchase felt like a one-time event — a necessity — but he quickly dug himself into a deep financial hole. (Fortunately, further digging was prevented by the financial companies canceling all of his credit lines.)

Jay's first solution was to switch to a card that required him to pay the entire balance each month. This led to some tough months, including last-minute scrambling to raise cash by selling CDs and books. It also brought his charging down to a manageable amount. Still, although he always scraped together just enough to stay out of debtor's prison, Jay never had a cent left over for the savings account that might someday become a down payment on the beach house he dreamed about.

That's when Jay's credit card company stepped in: it offered a new plan that would bill an extra amount to your credit card each month. This seems like the wrong kind of progress. How did having even more to pay help Jay save money? The trick was that the extra amount billed was invested in a mutual fund. This made the monthly adventure of paying off the card even more harrowing, but it worked. He always figured out a way (and little by little it came from charging less), and by doing so he accumulated $250 every month in savings.

One of the most effective savings mechanisms for us is to hide money. Who are we hiding it from? From ourselves or, more precisely, from that more impulsive part of ourselves. Jay was able to begin saving only by setting up a separate account that he never saw and that was extremely difficult to access.

If you have a job, you are already hiding some money from yourself in the form of social security. Although it is not technically a savings plan, social security helps us save for retirement. Essentially, the more we earn, the more the government will pay us each year when we are retired. For all of its well-documented flaws, social security has worked to ease poverty among older Americans. When the program was enacted, people over 65 were the poorest segment of the American population. Now they are the richest.

Another proven way to save, successful precisely because it doesn't feel like saving, is to buy property. Although the average sixty-year-old American has only $8,300 in financial assets, retirees have over $35,000 in the form of home equity. Failing to keep up the mortgage payments can result in losing the property, so even terrible savers turn out to be surprisingly successful at scraping together enough to avoid default.

In the 1980s, Brooke Shields made a series of racy advertisements for Calvin Klein jeans. In one she says, "When I get money, I buy Calvin Klein. If I have anything left over, I pay the rent." Successful savings techniques share a bit of this seemingly warped set of priorities.

Effective savers can say, "When I get money, I lock some up as savings. With the money that's left over, I purchase food and shelter." People save when the money comes out of income before other needs. As long as the amount of savings is fixed and required in the form of a mortgage payment or payroll deduction, most people find a way to make ends meet. If savings are simply whatever money is left over after buying, the result is usually no savings at all.

Setting up mechanisms for automatic savings can be incredibly painful, but nearly everyone gets over the pain and adjusts to their new income. Mortgages and secret mutual funds are just peachy for rich folks, but what about those of us who are hanging by a financial thread?

The trick is to pick the right time to increase the amount of hidden money. For example, when we get a raise, we can increase the contribution to our retirement account so that our take-home pay remains constant. We can whine all we want, but we know it's possible to live on the old salary because we actually did.

With growing government surpluses, it's also likely that we'll get a tax cut in the next few years. If so, that will be another excellent time to ratchet up the savings. Similarly, any windfalls such as tax rebates and gifts are best invested immediately.

The book *The Millionaire Next Door* describes the behavior of average folks who became wealthy. The surprising conclusion is that most people get rich because they spend less, not because they earn more than the average. Millionaires, for instance, hold off an extra year or two before trading in their decidedly non-exotic cars and are more likely to sport a Timex on their wrist than a Rolex.

⌒

Saving more. Why do we need so much help saving money while other behaviors come so easily? The answer is that we need little help learning behaviors that have been critical to human survival and reproduction for thousands of generations. We instinctively solve ancient problems, and it's only when our instincts fail us that we've got to buckle down and learn. For a dramatic example, consider how babies react to dangerous objects.

Place a loaded pistol in a playpen and the babies will play with it just like any other toy, giggle, and perhaps even place the gun in their mouth. In contrast, put a plastic snake into the playpen; the babies will cower in fear. Show a person of any age a snake — or even a picture of one — and you will elicit a dramatic response, including sweaty skin and an increased heart rate. It doesn't matter whether the person is in America, Europe, Japan, Australia, or Argentina, the response is the same. This is true even in Ireland, which has no native snakes.

Why do we have an instinctive fear of snakes and not of guns? In 1998, guns killed more than thirty thousand Americans; snakes killed fewer than two dozen people. In the United States, you are literally eight times as likely to be struck by lightning as killed by a snake. Nevertheless, snakes produce one of the strongest instinctual fear responses.

We ought to be very afraid of guns and relatively unconcerned by snakes, but we are built in just the opposite way. A bit of reflection resolves this puzzle. The genes that cause instinctual fear, like all genes, have been handed down to us from our ancestors. Snakes caused many human deaths when we lived as hunters and gatherers. In contrast, guns didn't kill a single person until very recently. Accordingly, we loathe our ancient enemy, the snake, and have no instinctual response to novel threats regardless of how deadly.

Other primates are also killed by snakes and have the same genetic hatred. Even adult chimpanzees and monkeys that have spent their whole lives in zoos and have never seen a snake share our instinctual herpetological fear. They become terrified and agitated immediately on seeing their first snake.

In contrast to our long evolutionary history with snakes and other animals, try to imagine the following conversation between your great-great-great- . . . great-grandparents as they sat around the campfire ten thousand years ago:

Husband: "Honey, I'm thinking of putting 25% of our savings into floating-rate Japanese bonds with an option to swap into Eurodollars. What do you think?"

Wife: "That's crazy. Over many thousands of generations, we've all learned that stocks are better investments because of their higher long-term return and more favorable tax treatment. All humans know to invest in technology firms. I hear that Fidelity has a new fund for investing in companies that can make fire."

Ridiculous. Our ancestors knew nothing of financial instruments. Accordingly, we are no more likely to have instincts to make arcane financial decisions than we are to fear guns. Our instincts for saving for the future simply aren't wired for modern financial choices. Maybe in a thousand generations but certainly not this fiscal year.

Our ancestors would obviously be confused by many modern financial choices. They'd even be ignorant of money, another modern invention. Let's consider how recently humans developed money.

The first ways to borrow and save used food as currency. In Lapland, all the way through the nineteenth century, people settled debts and secured housing for the winter with payments of cheese. Now, using cheese to pay the bills is not much of an improvement over Mother Nature. Elephant seals save via fat on their brisket while these Laplanders used fat-filled cheese in their baskets.

More recently, we have learned to amass more easily exchanged currencies. North American natives and settlers used the proto-currency known as wampum, purple and white beads made from shells. Moving south, cocoa beans were long preferred in Central America. Although they wouldn't last forever, they were easy to count, pleasant to handle, and you could always eat them in a pinch. Try doing that with a quarter.

And speaking of quarters, when exactly did we move beyond all of these essentially animal-like methods of storing value? When did we finally envision our modern concept of money?

The first minted coins appeared in the kingdom of Lydia, a center of international trade around modern Turkey and Greece, at the beginning of the seventh century B.C. The idea didn't exactly catch on like wildfire, though. As a Japanese proverb of the time states: "Wise rulers in all ages have valued cereals and despised money. No matter how much gold and silver one may possess, one cannot live for a single day on these. Rice is the one thing needful for life."

The difficulties of money are compounded at the intersection of cultures. Imagine the plight of the French singer Mademoiselle Zelie. In the course of a Pacific Ocean tour she played a concert in the Society Islands and received in payment her standard one-third of the box office.

Much to her chagrin, however, this amounted to three pigs, twenty-three turkeys, forty-four chickens, five thousand coconuts, and considerable quantities of bananas, lemons, and oranges — literally a third of what the box office collected. This would have been worth a considerable sum in Paris, but without an ark to get it home, it was virtually useless.

For the two and a half millennia that it has been possible to stockpile wealth in the form of coins, people have resisted. Old habits die hard, especially when they're in our genes. The chief drawback to money, of course, is that coins have value only to the extent that you can trust other people. Unlike the cocoa beans or rice, coins have no intrinsic value. As a consequence, right on up to the last century, we've got Laplanders assuring their skeptical landlords that "the cheese is in the mail."

Our brains are built by genes that excelled in a world without money. When it comes to padding our bodies with a bit of fat, we have powerful instincts. Similar instincts for minding our cash haven't had time to evolve.

If we take our snake experiment to the highlands of New Guinea, we have a hard time finding the same fear of snakes. Showing snakes or pictures of snakes amuses adult New Guineans. It doesn't frighten them. This seems a bit odd. Why the difference from nearly every other society tested? In New Guinea, unlike New York City, snakes abound and still kill many people. There is even one recorded case of a massive python killing and completely consuming a fourteen-year-old boy on a nearby Indonesian island.

If anyone should be terrified of snakes it would seem to be the New Guineans, who are still killed by them. Instead they laugh at our naive, generalized fear. Experience and learning are the explanation. From the time New Guineans are small children, they encounter snakes — only a third of which are poisonous — with great regularity. In the process, they learn to identify the nasty and the harmless snakes, often capturing the non-poisonous ones for eating.

The New Guinean naturalists have learned to modify our innate fear of snakes, capitalizing our big brains' ability to alter the program. Similarly, while babies show no innate fear of guns, people quickly learn the appropriate response. From these victories in modifying our instincts, we can gain inspiration for transforming our relationship with money.

Can people really change such firmly entrenched behaviors? Absolutely. The truth is, they're not even that firmly entrenched. It only feels as if we've been in debt forever. In fact, the number of people going bankrupt in the United States has changed by 300% since 1980. Although the change has been in the wrong direction, it shows we can change. The same applies to American savings behavior, which has moved steadily toward more spending in the last two decades.

Further evidence of our ability to save comes from other cultures with more frugal ways. Ironically, as hard as it is for Americans to boost their savings rate, the Japanese economy has stagnated from exactly the opposite problem: too little consumer spending. So while our instincts prime us to consume too much, Japanese frugality proves that those instincts are malleable enough to allow good savings behavior. Our real problem is that we are too flexible in our savings behavior. Companies know this weakness and attempt to manipulate us for their profit.

⤿

Buy me on credit! Financial firms make money the old-fashioned way, they charge high interest rates when they lend us money and pay low interest rates when we park our savings with them. In search of profits, the firms prey on our poorly honed financial instincts and exploit some of the quirks in our genetic legacy. Knowing their tricks can help us navigate the modern financial jungle.

Take Homer Simpson. He orders virtually every product that is advertised on TV. He sees someone with a shapely physique and immediately orders the twelve-cassette package that will teach him to lose weight. The cassettes are shipped immediately, while the money will not be due for ninety days or more. Homer's impulsive buying is funny because it is only slightly more impulsive than our own.

It's as though our brain can't quite grasp that money doesn't lose its value over time. As a matter of fact, that's exactly the problem; our brains were built for a world in which the currency of the day *did* lose value over time. Put simply: food rots. In that world, the savvy investor ought to devalue future payments severely. Unfortunately, our brain plays by yesterday's rules, so we are an easy mark.

Sure enough, companies do exploit our built-in impatience and often succeed best when appealing to our desire to have it all now. The ability to take home a fabulous washer-dryer set today, with no payments for sixty days, tickles the fancy of that little hunter-gatherer deep inside us. Never mind that in the end we will pay far more than we think is fair. At the time of purchase, our outdated instincts guide us in the proper balancing of value today against costs to be paid in the future.

The road to Hell, it is said, is paved with good intentions. We often waste money when we expect to change for the better but instead continue our impulsive behavior. In an investigation of good intentions, researchers studied people's willingness to watch serious movies.

In one group, subjects were asked to choose a movie to watch for that night. In the other group, subjects were asked to choose movies that they would watch on each of the next three nights. For this group, the movies would be viewed over the three days, but the choices were all made on the first day.

An interesting pattern emerged. When choosing for tonight, people in both groups selected lighthearted romances, comedies, and action films. When choosing for future evenings — tomorrow or the next night — people selected more serious films, such as *Schindler's List,* which portrays Nazi concentration camps, as well as films in foreign languages.

On day one people said, "I'll watch something fun tonight, but tomorrow I'll watch a film I ought to see." When tomorrow came, however, they again wanted to have fun and would have switched to *Groundhog Day* if possible.

Companies know that we are overly optimistic about our future behavior and use this knowledge to make money. For example, they offer us credit cards with a low introductory interest rate. The catch is that the interest rate will increase substantially after six months. There's nothing illegal about this. The banks don't even have to hide these terms in the fine print (though they always seem to). They could put them in neon lights on a billboard: YOU'LL PAY SUPER-HIGH INTEREST RATES, BUT NOT FOR SIX MONTHS.

Because of our relentless optimism that the future will be different — and better — than the past, we flock to these sorts of deals. (If aliens ever conquer Earth and keep humans as pets, they'll probably view this irrepressible optimism as our most endearing feature.)

When we sign up for such plans, we look forward to a new and improved us and expect to take the firms for a ride. We don't really care if they're going to charge us exorbitant interest rates in six months because we plan to be debt-free (and thinner) soon. When the six months end, though, we are usually still saddled with our debts. One result is that the average American torches a fifth of his or her income on credit card payments that are mostly just interest.

Taking control of our finances. How can we prosper among the loan sharks and a sea of tempting offers designed to stimulate and exploit our desires? Well, we can't rely on our instincts. Instead, we need to continuously hone our financial training. We have to turn the tables on banks and businesses by doing to them exactly what they do to us. Remember, they make money by charging high interest rates on the money we borrow and paying low interest rates on the money we save.

Step 1 in turning the tables is to think of everything in the same currency, namely after-tax interest rates. Regardless of whether something is called a loan application fee, an interest charge, or a balloon payment, all that matters is the interest rate. A variety of excellent books and software programs exist to help us convert costs of all types into an effective interest rate. With a concerted effort, anyone can do the required calculations. As obvious as this may seem, many of us don't even know what interest rates we are currently being charged.

Step 2 is organizing your financial house. Whatever debt we have should be at the lowest possible interest rate — and tax deductible if possible. Similarly, whatever savings we can accumulate should produce the highest possible after-tax return.

For example, if we owe $3,000 on our credit cards and have $2,000 in a savings account, we are giving money away. Savings accounts pay paltry interest rates, in the 2%–3% range at best, while credit cards charge a rate closer to 20%. We can act like Wall Street bankers by using some of the savings to decrease the credit balance. Money that was earning us 2% instead will now save us the 20% we are being charged.

At one level this need to organize and rationalize our finances is obvious, but at a deeper level it requires us to suspend our instinctual response to tempting offers. This is all the more difficult because firms are always designing, consciously or otherwise, programs to fool our intuitive concepts of value.

Step 3 is to be realistic about our own behavior. This may be the hardest thing of all when it comes to managing money. Even though we think we'll do better tomorrow, the best predictor of our ability to rein in future desires is our past behavior. We shouldn't expect to sit through three hours of an obscure documentary when *Austin Powers* is on cable. Similarly, we shouldn't accept financial offers that will save us money only if we become completely different people.

Finally, take advantage of firms. Although each company is trying to gouge us as much as possible, they are competing for our business and may be forced to offer great deals. For example, a phone company may have an introductory low-price deal or some sort of signing bonus. It's betting that we'll stick with it when the offer ends. However, we can take the freebie and move on to another offer before the deal gets worse for us. Companies hate consumers who jump from freebie to freebie, but it's legal and profitable.

Evolution has produced elegant solutions to many problems. If we stumble physically, our bodies' systems react instantly to catch us or minimize the pain of the fall. There are no instinctual protections in the financial area. With every major decision, we must suppress our gut response and use our learned financial tools to make the best choices.

Fat Please don't feed the humans

Furry love handles. Chantek is a smart, lovable orangutan who lives at the Atlanta zoo. Trained in sign language, he has a vocabulary of more than 150 words and is considered a decent artist. Now in his twenties, he was born at the Yerkes Primate Center in Atlanta and then spent nine years being raised as a human — complete with diapers and infant formula.

Growing up in this human setting, Chantek became *really fat*, weighing in at five hundred pounds, roughly three times his ideal size. Afraid that the massive bulk would collapse his lungs, scientists placed him on a strict diet. Formerly five hundred pounds of fun, he became four hundred pounds of anger. During the diet, his favorite sign language symbol became "candy." He refused to draw and instead ate the crayons given for his artistic use.

While on his diet, Chantek even pulled off an escape. He threatened and could have easily killed a janitor, but chose instead to attack a 55-gallon drum of food. He was eventually found sitting next to the up-ended food barrel, using all four limbs to stuff monkey chow into his mouth.

Chantek is unique, not only for his human contact and his linguistic and artistic abilities but also for his weight. You see, there are no fat orangutans outside zoos and research centers. Wild orangutans, despite sharing Chantek's genetic zest for a fine meal, maintain a svelte 160 pounds or so because food is relatively scarce and difficult to obtain in the jungles of Borneo.

Like Chantek, many of us have trouble staying skinny and healthy. As we'll see, easy living with plentiful food is the source of weight control problems for humans and captive orangutans alike. Our appetites were built in a world where plentiful food was inconceivable.

Really poor people are still baffled by the idea that overeating can be a problem. On a trip to the East African country of Uganda, Terry attempted to explain bulimia to a group of women. He started by saying that bulimic people purposely vomit after eating. "What is wrong with the food?" the Ugandan women asked. Nothing, said Terry. Bulimics just want to get rid of it. The women stared blankly at him, clearly having trouble processing this information.

After several more attempts to comprehend the disorder, the Ugandan women left, convinced they had just witnessed either a translation error or a Westerner's odd joke. How do you convey the problem of too much food to people who are chronically hungry?

In many poor countries, fat is still a sign of wealth and the word "prosperous" is used to describe heavy people. Nigerian brides eat and relax in "fat rooms" to put on weight before their weddings. Plumper wives are more pleasing to their husbands, and their energy reserves are useful for pregnancy.

Outside the industrialized countries, famine and malnutrition are still common, with half of the developing nations experiencing food shortages in a typical year. Under these conditions, it pays to build up some reserve against the hunger season that often lurks ahead. Indeed, our nearly insatiable appetite was once a survival feature of human biology. A profound love of food helps people to pack on a few extra pounds and thereby survive periods when food is scarce.

Those thrifty genes still drive our behavior. Holdovers from the uncertain times of our ancestors, they function as though our world has not changed. It has. In our zoo-like environment we have continual access to food, and a suburban famine seems to occur when dinner is delayed for an hour or two.

Our ancestors lived off the land by hunting animals and gathering plants. To understand how different our world is, consider the life of people who forage for survival even today. To acquire food, they expend hundreds of calories each day walking and then spend hours preparing meals. Just staying alive requires lots of energy — energy that can be found only in food.

For those of us in industrialized societies, a few taps on the accelerator take us to supermarkets brimming with food ready to be cooked or eaten. The garage is only steps from the kitchen, and the supermarket has a parking lot that brings us to within fifty feet of the food. If driving to a market is too taxing, we can telephone for pizza or Chinese food.

Our lives are filled with machines — remote controls, phones, refrigerators, electric can openers, TVs, computers, and cars — all of which help us get our fill of calories, social contact, and entertainment with minimal effort. How many steps did you actually walk today? For most of us, the answer is "very few."

Sitting on our couches, sitting in our cars, sitting at our desks, we are not experiencing any sort of energy crisis. Most of us already have too much stored energy on our bodies in the form of love handles, saddlebags, beer bellies, and other unwanted lumps of flesh.

Powerful, instinctual hunger kept our ancestors going in a tough, energetically demanding world. Imagine a time when the individuals of a population vary in their appetites. One gluttonous type thinks of food day and night. Another type becomes satiated once their daily needs are met. Of these types, who has the biggest surplus of energy stored in their thighs and buttocks when food is scarce? Who weathers the famine with calories left over for reproducing? Who is most likely to be your ancestor? Fatties, fatties, and fatties again.

This hunger was a survival-enhancing feature in our genetic programming. Now it is a bug in that programming. The consequence of our perpetual hunger is not news: one of every four Americans is obese. In terms of size, plumpness gets labeled obesity when our body mass index (weight in kilograms divided by height in meters squared) hits 30. (Have they purposely defined it in a way that will be meaningless to most Americans?) This translates to about 209 pounds if you are 5'10" and 180 pounds if you are 5'5".

Predictably, we keep gaining weight, both as we get richer as a society and as we age individually. Most of us would reduce our risk of heart disease, stroke, and diabetes if we lost even as little as ten pounds. We know this. That's why so many of us are trying to lose weight — and the rest are eating scared.

The caged Chantek is fat because his genes are adapted for the wild, where food is scarce and life hard. Our human ancestors lived in conditions more similar to the Indonesian jungles where wild orangutans roam than to modern industrialized circumstances. Presumably obesity was as rare for ancestral humans as it is for wild primates today. Just as dogs and cats often get chubby around the house, zoos are populated with animals that have weight problems. We would be better off if we wore signs that read, PLEASE DON'T FEED THE HUMANS.

⌐

Hunger pains. Whatever our genetic endowment, any one of us will lose weight if we expend more calories than we consume. The equation holds, whether the calories are consumed in hamburgers or in fresh vegetables. Similarly, it doesn't matter whether energy is expended in the weight room or the bedroom. We can't get off the hook by simply pointing out that our thrifty genes are making us fat.

An important dieting concept is "weight change = calories in minus calories out," but this is only part of the story. The challenge lies in making lasting behavioral changes that reduce input and increase output. Almost anyone can torture themself with food deprivation for a while. Short-term starvation is even the fastest way to take off a few pounds before that dreaded high school reunion. Permanent weight loss, however, is much more elusive.

Living in a perpetual state of hunger, though, may be the worst plan for permanent weight loss. At least eight human guinea pigs can attest to this. In the Biosphere 2 project, eight people entered a self-contained 3.2-acre world filled with plants and animals. This Biodome project was designed to explore sustainable living with minimal impact on the environment, but it may have ended up telling us more about human nature.

Although the scientists knew it would be difficult to live in the Biodome's fishbowl world, the greatest hardship they endured came as a surprise. One of the key Biodome findings? Hungry people are grumpy. The Biodomers lost weight because they had very little food. This shortage was partly a planned experiment on the effects of a low-calorie diet and partly caused by crop failures. As predicted, the weight loss caused health improvements, including a reduced risk of heart attacks.

On their sparse diets, the Biodomers also argued constantly, got into ugly food spats, and frequently squabbled over dinner portions. After leaving what they dubbed "the hunger dome," one of the eight said, "If we ever all start talking to each other, that would be a major accomplishment." When the participants resumed their normal lives, they all returned to their previous weights.

You don't need to spend $100 million on a Biodome to learn this lesson. Common sense and scientific experiments reveal the pain associated with hunger. In one study, people were kept hungry for six months. Over time, increasing hours were filled by fatty food fantasies. Recipes even displaced sex as the favorite topic of discussion.

So we can't simply stay hungry. But are there other ways to change our eating habits to achieve permanent weight loss? To find out, researchers put a group of monkeys on a very low calorie diet. The monkeys shed pounds initially, then stabilized at a low weight for a full two years. Although their simian minds may have been filled with monkey chow fantasies, their behavior seemed normal to human observers.

After two years these lean monkeys were given unlimited access to food. Did they stay skinny? Absolutely not. After spending close to 10% of their lives at a constant, low weight, they quickly returned to their original weights.

A variety of human and animal studies point to the idea of a "set point" for weight. Just like a thermostat, when weight is below the set point, the body sets out in search of calories, and when weight is above the set point, the mind and body are free to pursue other goals.

If the set point is true, how does it work? One method is by changing the metabolic rate. Dieters often complain that their metabolism slows down and they gain weight "just by looking at a piece of pie." Recent experiments validate this folk wisdom.

In one study, people were put on diets where they gained or lost 10% of their body weight. After several months at this new weight, doctors measured their metabolic rates. People who had lost weight did indeed have lower metabolic rates. In addition to having fewer pounds, their bodies had become more efficient in maintaining each of those pounds. Conversely, the metabolic rates rose in those participants who gained weight.

In addition to changing metabolism, the body combats dieting by releasing chemicals to induce eating. In particular, a component known as neuropeptide Y (NPY) causes carbohydrate cravings to go through the roof. NPY production increases with weight loss, and it gets pumped out at spectacularly high levels when people severely restrict their caloric intake. In other words, starvation-like behavior sends an alarm throughout the body saying, "We're in trouble. Eat anything and everything in sight."

These metabolic response systems frustrate our dieting efforts, but they are crucial to people without secure food sources. Along the way to begetting us, our ancestors lived through famines and repeated bouts of hunger. They sought food aggressively, and they sought it more aggressively in times of scarcity. They stored as much of that food energy as possible to prepare for lean times, and without refrigerators, the best way to store it was in body fat. Simultaneously, they evolved mechanisms to cut back metabolism during the hungry times.

Severe dieting may not just be fruitless, it may also be bad for our health. As we saw, dieting slashes the metabolic rate, damaging crucial functions. Just as a family facing a financial crunch may defer important work — such as fixing the car's brakes — a starving body slows down a variety of systems or shuts them off completely. Hungry lab animals almost completely lose their sex drive and may be less adept at fighting infection.

For all we have learned, important questions remain. Can the set point be changed or will our mean genes always win the weight loss battle?

Losing weight. Pharmaceutical companies have been looking for a magic pill, and along the way, they've made one almost absurd discovery. In clinical studies, researchers testing new diet drugs always have a placebo group. Some patients get the test drug, others get pills without any active ingredients, placebos. Because the placebos look exactly like the real pills, no one, not even the doctors, knows which patients are getting the drug candidate. The goal is to separate the effects of the new drug from the testing process.

As expected, some of the new drugs work and others fail, but here's the strangest finding: people in the placebo groups always lose weight. In a study on the effectiveness of the new diet pill Xenical, for instance, more than 25% in the placebo group lost at least ten pounds. How can I score some of that placebo? Is there something magical about it? Of course not. But here's the trick: while those who take the placebo aren't using drugs, they *are* keeping track of their weight and are more aware of what they eat than usual.

So perhaps it's just an awareness of eating and the keeping of records that helps those in the placebo group lose weight. This may also be the "secret" behind the success of some crazy, unscientific diets that advocate, for example, eating only food of a certain color (but "as much as you want") on a certain day. Careful monitoring is a crucial component of gaining control.

A similar result comes from another line of research. While recognizing that most dieters fail to keep weight off, one study interviewed people who had succeeded in long-term weight loss. One behavior that these success stories had in common was monitoring eating habits without severe dieting. These winners don't starve themselves, but like the placebo group in drug studies, they are continually vigilant about what goes in their mouths.

Beyond keeping track, another simple precaution can help. Consider this: On the way to a summer barbecue, Jay forced himself to eat three plain bagels. He knew there would be all sorts of tasty but super high–calorie foods at the dinner. He especially feared the cheeseburgers and nachos he knew he'd want. By eating the bagels as a sort of preemptive strike, he decreased his appetite with a minimum of fat calories and had much more willpower when tempted by dangerous delicacies.

Successful weight loss requires planning what sorts of food you want to eat and following that plan. When we try to reduce the *number* of calories we eat, our genetic systems fight us every step of the way. We can, however, much more easily win the battle over the *type* of calories we consume. By eating boring bagels, Jay followed his plan to eat only low-fat foods. This ability to choose the type of calories we consume may seem minor, but it can be central to controlling just how many calories we actually eat.

So we've got at least two steps — albeit low-tech steps — that can work for us. The first is to decide on what sorts of foods to eat. Whether we want a low-fat, high-carbohydrate diet or the Atkins low-carbohydrate diet, it's crucial to decide in advance. The second is to keep records of our goals and honestly document our consumption. Having to admit to three or four chocolate chip cookies — even on a list that no one else will see — proves to be enough to help many of us resist the temptation.

A few people will overcome their evolutionary systems and rapidly starve themselves to Ally McBeal–dom. For the rest of us, our relentless food-seeking genes will, sooner or later, induce us to eat about as many calories as we have eaten for most of our adult lives. Knowing this, we can plan our next meal, be realistic, and enjoy our food more.

Does any of the following sound familiar? You wake up in the middle of the night thinking about those cookies stashed in the cupboard. You get up and eat them all. Or you go to the supermarket resolved to buy only healthful foods, then buy a chocolate bar and eat it in the parking lot. If you have had these problems, you are human and you have normal genes.

Socrates used to say that he was the smartest man in Athens because he knew he was really dumb. When it comes to controlling our diets, strength requires knowing that we will be weak. Recognizing our future weakness allows us to minimize how frequently we will fail and to limit the damage done when we are feeling weakest.

Let's return to that late-night bingeing. Each one of us knows which foods prompt our nocturnal munchings. Maybe it's Twix candy bars. Maybe it's bowl after bowl of breakfast cereal. But there's usually a period, after dinner perhaps, when those Twix bars or Ding-Dongs have no appeal at all. In fact, we feel so satisfied that we can't imagine ever craving another Ding-Dong.

But the "you" that wakes up in the middle of the night has seriously different ideas about those treats. Defeat that monster within by preemptively building a fence. Throw out the Ding-Dongs right after dinner — or better yet, don't buy them. Leave a note in the empty cupboard: "Dear meaney geney monster, Ha! There are no Ding-Dongs. Eat a rice cake and thank me in the morning."

Each of us has fairly predictable periods of strength and weakness, so we should take preemptive steps when we are strong. While the exact problem and solution will be different for each person, this theme is constant. Here are some problems and solutions that work for some people.

Problem: I like to indulge my passions for certain junk foods, but I overeat. For example, I decide to eat some potato chips and buy a big bag, planning to eat just half, but then eat the whole bag.

Solution: Open the bag of chips and divide them into two piles. One you will eat and the other you won't. Destroy the chips you don't want to eat. The destruction must be performed before beginning to eat. It's not that hard when you know you're about to eat the entire other tasty pile. This is your moment of strength. When you throw them away, be sure to make them inedible so that the monster within you won't be picking through the trash at four in the morning.

Problem: I plan to eat nothing between lunch and dinner. But in the afternoon I often become very hungry and eat chocolate.

Solution: From your overall dietary plan, choose an appropriate afternoon snack. Make sure to take it with you every afternoon. When hunger strikes, you already have the correct food. It's unrealistic to believe that you will be able to stay hungry all the time. You have to eat and are going to eat. Accept this fact but make it easier to eat the foods you choose.

Problem: I buy the wrong food at the supermarket. As soon as I enter the store, my cart moves, almost of its own free will, into the aisle with the soda and chips in spite of my vow, just moments before, to buy only healthful food.

Solution: One well-known option is to shop only after eating. If this works for you, be sure to eat before each trip. If not, take more drastic steps. For example, make a list and send someone else to shop. This is becoming easier nowadays with Internet grocery shopping services; just don't punch out the delivery person for bringing brown rice instead of brownies.

Problem: When I host social events, I stock foods outside my planned diet. I don't mind eating off my plan a bit during the party, but afterward I often gorge myself on the leftovers.

Solution: As soon as the party ends, get rid of the "dangerous" items. Send them home with guests or give them to your neighbors. If all else fails, bury them in the backyard. Ignore your inner voice screaming that it is wrong to destroy food. Of course, it is always preferable to avoid buying it in the first place or to give it away. When these options aren't available, however, nothing is gained by eating something you'll wish you hadn't.

Problem: When I'm on a plane, they serve me a meal that includes a tasty brownie for dessert. I'm a prisoner on the plane and bored and tired and hungry, so I usually eat it.

Solution: Fortunately, with your sandwich and brownie they usually include a small packet of mayonnaise. When Jay receives his airline meal, he promptly opens the mayonnaise and smears it all over the brownie. This way, he's not tempted anymore. He is actually repulsed by it. Who knows? In the future, we may carry a bottle of *Mean Genes* Anti-Condiment Spray in flavors such as Mildew, Rotten Eggs, and Fish Entrails.

All these stories reflect Socrates's insight. Knowing we will be weak allows us to be strong.

Survival of the laziest? In 1984 Peter Maher weighed more than 250 pounds and smoked three packs of cigarettes a day. Although he had no athletic history, he wagered some of his mates (presumably while hefting a pint or two) that he could complete a marathon. He began running and won his bet.

Along the way he discovered that he possessed a tremendous natural talent. He became a full-time runner and has clocked a 2:11 marathon, just 6 minutes slower than the world record. And in the process he dropped to a gaunt 140 pounds on his 6-foot, 4-inch frame. He still worries about his weight, but now he fears becoming too skinny.

Even if we can't run marathons, exercise has obvious benefits in our battle to become and remain thin. Physical activity takes energy, increases metabolic rate, and makes bodies more muscular. Peter Maher and a lucky bunch of people love exercise. For most of us, however, slapping on the Nikes and pounding out a few miles ranks right up there with a visit to the dentist. Why do we have so much trouble doing something with such obvious benefits?

Most animals are lazy. Take mice, for example. Scientists investigating the effects of extreme exercise naturally want mice that are long-distance runners. The trouble is, mice don't want to run marathons any more than most humans.

When mice are placed on tiny treadmills, many of them will simply go on strike. They will even sit on the moving belt to the point that the skin on their butts begins to get scratched and scraped. The mice are ingenious in their ability to avoid exercise, positioning themselves against walls, splaying their feet at extreme angles — anything to avoid jogging. Any one of us who has made elaborate excuses to escape a workout can empathize.

In addition to orangutans like Chantek, the Yerkes Center has a large number of chimpanzees. Predictably, this population of zoo primates is also overweight and lazy. One of the matriarchs is a female called Natasha, known because of her bulk as Na-"tank"-a.

Even though the chimps are very well fed, most of them still excitedly yell and run around if a person shows up with a box of oranges. Natanka is not so easily moved. She plants herself directly below the fruit platform and begs with subtle arm movements that require no more energy than changing the TV channel with a remote control. She will move only about six inches to snatch a ripe fruit; getting her to travel across the compound requires dozens of oranges, each artfully thrown a few inches ahead of her.

Laziness is good for most animals. (Try not to let your kids find out about this.) To understand this, we have to leave our couches and think like wild primates. Energy in the form of food is hard to obtain, and once gotten, not to be squandered.

This is why lions sleep most of the day, mice squat on treadmills, and people avoid the gym like the plague. The only people who exercise in poor countries are the privileged — the rich, the tourist, or the full-time athlete. Similarly, modern foragers lead active lives, but the concept of unnecessary exercise puzzles them.

Evolution favors the frugal and casts a hard, wary glance on an organism that frivolously wastes energy. What happens to animals that expend energy needlessly? They die, and their genes die with them. We descended from humans who were frugal with their physical activity, and we carry their energy-conserving genes.

∽

Getting off the couch. Exercise is good for us, but evolution has built us to love laziness. Our genes still think there is a famine around every corner and hoard every calorie by inducing lethargy whenever they can. Can we ever get ourselves off the couch?

Let's rejoin our mice who hate running. While it's hard to get them to jog frivolously, they do love a good run under the right conditions. For example, if they are hungry, they spend a good part of the day running. Why? Well, among other things they are looking for food. Of course they don't make any progress turning a wheel in the lab, but they think they are covering territory looking for food. In one study, hungry mice ran about three miles every day, roughly twice the distance of well-fed mice.

In a related and slightly diabolical study, a scientist rigged up a mouse food-for-exercise incentive scheme. The mice could only obtain food if they turned an exercise wheel: they were rewarded with a small pellet of food each time they ran a certain number of revolutions (anywhere from 75 to 275 revolutions per pellet).

The more revolutions they were required to run for a food pellet, the more they ran, until the most active mice were running for an average of 10 hours a day. Not surprisingly, these animals ended up weighing far less than those that didn't have to work for their food. (Perhaps Richard Simmons should design a *Mean Genes* television or refrigerator that can be powered only by an exercise bike.)

In our quest to be more active, we can perpetually fight against our genes or we can outsmart them. If we can set up situations in which activity is required to meet a worthy goal, the mental energy and willpower required can be very low. The type of goals that our genes consider worthy varies from person to person, but they have some common themes.

Anita, a twenty-six-year-old woman who lives in Boston, has a walking partner for her brisk daily 6:00 A.M. stroll along the Charles River. On mornings when Anita is particularly tired, she naturally wants to call her friend to cancel, but she is thwarted. Her walking partner has roommates and, with only a centrally located phone, any call will disturb them. For Anita, the combination of not wanting to awaken her friend's roommates or miss a planned appointment is highly effective. For many people, a workout partner or a team sport makes exercise much more palatable.

Money can motivate others. In one study, psychologists approached a group of people in line to buy season tickets to the theater. They gave half the theater buffs a reduced rate on tickets while the others paid full price. At the end of the season, it turned out that the people who had paid more for their tickets attended significantly more plays. Similarly, some people find paying for a health club membership pushes them to more frequent exercise. The desire not to waste money can be stronger than the drive of sloth.

There is a related message in the latest episode about Chantek the orangutan. When we met him, he was dieting, hungry, angry, and dreaming of candy. After his escape and the monkey chow episode, he was moved to an new area with a much larger domain — several acres — and he must walk a bit to get his food.

Furthermore, because wild male orangutans are territorial and spend much of their time patrolling their part of the jungle, Chantek likes (or feels compelled) to walk around to make sure that no males are intruding on his turf. Of course in the zoo there will never be intruders, but his genes don't know that. As a result, Chantek is much more active and, even though he is no longer on a strict diet, he has lost fully *half* of his 500 pounds.

Nature abhors wasteful energy expenditure but will induce us to work under many conditions. Chantek has become more active because he has to work for his food and he likes patrolling his territory. Those frequently sleepy lions will sprint if there is a gazelle to be chased or a hyena to attack. If we can set up situations with appropriate rewards, we can similarly shed the cloak of laziness. By structuring our lives so that we must be more active, we can reduce our weight a bit and can be healthy without starving ourselves.

Food substitutes, surgery, and pills. We live in a zoo-like environment with excess food, surrounded by labor-saving machines that ease our every task and ensure that no desire is more than a button push away. Our genes have built us to love food and hate exercise; accordingly, the root of our weight problem is that our wild genes now live in a tame world. Those genes aren't going to change anytime soon, and there is no imminent return to food scarcity. Technology has gotten us into these problems by making us rich. Can it also inspire inventions to keep us thin?

One technological effort is development of food substitutes. To understand them, let's focus on the fat substitute Olestra for a moment. We love fatty foods because our tongues have thousands of specific detectors — taste buds — that stimulate our brains when we eat foods like nuts, avocados, cheese, and red meat. With this system, a fatty meal produces a bit of a brain buzz. This structure evolved because fat has the most calories per serving. Our ancestral genes reward us whenever we find calories; in this quest for energy, fat deserves — and receives — the biggest reward.

While we love our deep-fried nachos with cheese and guacamole, most of us hate the calories. Olestra is an attempt to give us pleasure without the cost. Specifically, it turns on those fat detectors but yields no calories at all. One part of Olestra has a chemical structure just like fat, so the detectors tell the brain to party, but the molecule is designed to avoid digestion. The mouths says "ahhh," but the gut fails to extract any calories. The result is a substance that deceives our bodies into feeling as though they have just had a satisfying meal.

Food companies use a variety of other substances, including Nutrasweet, that are similarly designed to fool our bodies. Fat, sugar, and salt all taste great but can be bad for us. The artificial molecules, on the other hand, hold the promise of toll-free satisfaction. In particular, these compounds and many others under development allow foods to be tasty and low in calories. It sounds simple. Does it work?

To find out, researchers secretly gave one group of people sugar cookies. A second group got cookies that looked the same, and tasted pretty good but were made with Nutrasweet. The researchers then noted how many cookies were eaten. *Voilà!* Both groups ate the same number of cookies, therefore those with real sugar consumed many more calories. A victory for technology? Maybe not.

The subjects were also asked to keep diaries of all their eating on the days around the cookie-fest. Those in the Nutrasweet group ate more than those who ate the sugar cookies. So much more, in fact, that the total caloric intake of the two groups was identical. Moreover, those in the Nutrasweet group preferentially ate more sugar. In the end, the people in both groups ate a lot of sugar and calories, but only some had also ingested a wad of Nutrasweet.

Current food substitutes aren't going to save the day, so some people are taking more drastic measures, such as a variety of surgical procedures. One option for people who have significant health risks from their weight is a process called stomach banding, in which a surgeon makes the stomach smaller. Patients with stomach bands get full and stop eating much more quickly than before the surgery. Patients in one group with banded stomachs each lost — and kept off for at least two years — about seventy pounds.

Operations that remove part or all of the small intestine also reduce weight. This shorter digestive system causes food to pass through before all of the calories can be absorbed. The overall effect is similar to eating Olestra: the mouth is tricked into happiness by tastes that promise calories while the gut is short-changed.

Somewhat less extreme is liposuction, the removal of fat cells. It is fast becoming one of the most popular operations in this country, with one hundred seventy thousand performed in 1998 alone. Unfortunately, over time patients tend to regain all their lost weight, albeit in different places. Given this result, liposuction may better be thought of as "body contouring."

If food substitutes and surgery can't guarantee permanent weight loss, what about prescription drugs? While there have been no unqualified breakthroughs in this area, the notion of building drugs that modify the body's weight system is sound.

During the two decades before it was shown to seriously damage the heart valves, for instance, the diet drug combination Fen-Phen was used by as many as five million American women. It combined an appetite suppressant with an amphetamine-like drug; like all successful weight-loss drugs, it interfered with our instinctual systems that are always seeking to acquire and convert food into fat reserves.

Another strategy for tinkering with the genetic machinery is simply to turn up the metabolic rate. A variety of products including Metabolife claim to increase the expenditure of energy without an offsetting increase in appetite. This approach is theoretically sound, but there are limited data on the safety or efficacy of these unregulated products. Clinical trials do, however, show that the stimulants ephedrine and caffeine can produce weight loss in the range of five to ten pounds.

The latest drug to hit the market is Xenical. Marketed under the name Orlistat, this pricey drug interferes with fat digestion by causing some of the fat in, say, salad dressing or an olive to pass through the body (sometimes distressingly quickly) without absorption. In clinical studies, Xenical has succeeded in helping people lose about ten pounds over the course of a year. The second year, these people regained some of their weight but still ended up somewhat lighter.

Current diet drugs can thus help us lose some weight, and the future is getting brighter. The ten pounds that Xenical melts from the average person may not make it a wonder drug, but it is a significant start. And for many people, those ten pounds are all they need to shed. As long as we live in our zoo-like environment of plenty, we're going to struggle with natural systems that seek out and store calories. But as we dissect the genetic machinery that makes and keeps us fat, the prospects grow for more effective drugs with fewer side effects to start tipping the scales our way.

CONSTANT **CRAVINGS**

Drugs Hijacking the pleasure pathway

The lure of drugs. John Daly says he's finally given up his efforts to quit drinking. Once one of the most promising young athletes in America, the professional golfer recently chose to return a three-million-dollar endorsement from the leading golf club manufacturer because it required him to abstain from alcohol. Daly said that trying to *stay sober* "had taken over my life and I was miserable." He blames a strong genetic desire for alcohol, and although he's sad about the many costs of drinking, he says, "It's great to be free."

John Daly is not alone. Rock stars fall with such regularity that it's practically front-page news when one of them doesn't have a drug problem. The allure of these little chemicals is tremendous, and although tabloid tragedy coverage is limited to the likes of Janis Joplin, John Belushi, and River Phoenix, millions of us face — and frequently lose — self-control battles with drugs.

Drug use is an everyday feature of modern life. Alcohol is omnipresent, and as a result, tens of millions of us endure everything from decreased job performance to liver damage to full-blown alcoholism. Alcohol is involved in 75% of spousal abuse cases. More than fifty million Americans smoke cigarettes, contributing to almost half a million deaths per year — more than a dozen times the annual number killed in traffic accidents. The litany of drug damage goes on and on.

Tiny chemicals have equally powerful effects on animals. When an ovulating female boar is exposed to a pheromone from a male boar's saliva, for example, she becomes immediately and completely paralyzed in a spread-legged mating posture. If you put rats in a cage with unlimited access to both food and cocaine, what happens? The rats consume the cocaine ravenously, ignore the food completely, and in short order starve to death.

This widespread love affair with drugs prompts a perplexing question: Shouldn't evolution produce industrious organisms, not drug addicts? To understand, we need to step back and think about the evolution of feelings. Why do our bodies have the capacity for pain and pleasure? Once we answer this question, we will see why we are so strongly drawn to dangerous substances like alcohol and cocaine. But first, let's begin the way we begin most journeys — with a cup of coffee.

Why is caffeine so damn good? David Letterman says, "If it weren't for coffee, I'd have no discernible personality at all." Indeed, caffeine is perhaps the most frequently used drug of all. Worldwide, more tea is consumed every day than any beverage other than water. Close behind is coffee. In the United States, 90% of the soda we drink contains caffeine. The average American drinks about a hundred gallons a year of these three beverages.

From philosophers and writers to scientists and musicians, coffee has been revered as a necessity for stimulating the creative juices. In his *Coffee Cantata* of 1732, J. S. Bach wrote, "Ah! How sweet coffee tastes! Lovelier than a thousand kisses, sweeter far than muscatel wine!" Two hundred years later, coffee's appeal had, if anything, grown. Isak Dinesen, whose autobiography was made into the movie *Out of Africa,* wrote, "Coffee . . . is to the body what the word of the Lord is to the soul."

The centuries of strong praise are well founded. Caffeine has powerful effects on nearly every animal species. Take rats, for instance. While all rats can eventually be trained to race through mazes, some learn quickly while others languish in remedial maze-running classes. Still, what they all have in common is that when they are given a caffeine pick-me-up before their maze lessons, they learn the solutions faster and remember them better.

Competitive bicycle racers have taken these results to heart. They have discovered that they can pedal 20% longer if they consume caffeine an hour before racing. Perhaps blurring the line between savvy training and competition gone mad, some insert caffeine suppositories prior to races to enjoy a time-released kick in the ass.

Given caffeine's ability to rev up minds and bodies, it's no surprise that sperm, too, do a little caffeine dance. Following exposure to extreme doses, sperm swim faster and wiggle more vigorously, increasing their ability to ford even the most viscous of cervical streams in search of a fertile egg.

Amazingly, caffeine seems to be safe for most people. Despite considerable searching for ill effects, there is no evidence that moderate consumption of caffeine increases our risk of anything beyond occasional jitters. For healthy people, there appears to be no increased risk of heart, lung, or kidney disease or even cancer.

How does caffeine work its magic? As long as we are awake, our brains are working hard. Our senses soak up data from the world around us: our sweater feels scratchy on our skin, the sun shining in the window is awfully bright, the kids are demanding attention, and our boss is yelling about some overdue report. And so on. All of this information is reported to the brain by special cells called neurons.

With each bit of information we process, millions of neurons are active. The problem is, much as a running motor generates exhaust fumes, all of this neural activity leads to a serious buildup of cellular waste products. Eventually our cells need a nap. Neuron "exhaust" takes the form of molecules, including one called adenosine. Prompted by the adenosine buildup, our bodies nudge us into bed when our batteries need recharging.

Anyone who has struggled to stay awake while driving knows the relentless pressure adenosine exerts on us to stop and sleep. Adenosine itself doesn't cause the sleepiness; it's a messenger that simply signals the surrounding cells to settle down. Caffeine blocks this sleepy message. Here's how.

Our brain cells communicate by passing chemical messages like adenosine to one another. Messenger molecules are teamed up with specific listeners — called receptors — on other cells. Adenosine and its receptors are exquisitely matched, like tiny locks that open only with the proper miniature key. When a cell releases adenosine, it fills adenosine receptors on nearby cells, passing along the message to sleep.

As the production of adenosine continues throughout our day, more and more receptors are filled. Increasingly our brain cells become sluggish, regardless of how strongly they are stimulated. We become tired. As we sleep, the night shift sweeps the adenosine away. When we wake, we feel better because we literally are more clear-headed.

But let's say we don't have the luxury of climbing into bed when we feel tired. Instead, we reach for a soda or a double espresso. The caffeine we ingest makes a beeline for our brains, and once there, it bubbles around between the cells. Here, because of a chance similarity in shape, the caffeine slips into some of those receptors intended for adenosine.

Once nestled in these slots, caffeine just camps out, blocking adenosine from doing its job. So perhaps we've been up for hours, working like crazy, and our brains are awash in adenosine. We should be dead tired. But with many of the receptors blocked by caffeine, the adenosine can't pass on the message that we ought to go to bed. Instead, we feel surprisingly alert and still ready to take on the world.

Caffeine keeps us awake because it interrupts the normal sleep signaling system. Some drugs obstruct the body's natural signals while others amplify messages. In every case, however, drugs masquerade as naturally occurring compounds, tricking our brains. Let's look a bit closer at just how dramatically all of these chemical messengers can influence how we see, feel, and experience the world.

If it feels good, do it again. Can the actions of a few brain cells really influence our mood or behavior? In the 1950s, a psychologist surgically implanted electrodes in rats' brains and stimulated them. Usually, sending a tiny current of electricity elicited little response. Positioning the buzzing electrodes near a part of the brain called the hypothalamus, however, seemed to make the rats happy. Actually that's the understatement of the year. Stimulating the hypothalamus made them ecstatic.

Subsequent experiments have shown that if the electric reward is doled out only when the rat accomplishes something — learns to navigate part of a maze, for example — the rat works at the task eagerly until it is mastered, craving the reward. As long as the rewards keep coming, the little rodents keep working, even to the point of mastering complex mazes that humans would find nearly impossible.

It's not the learning that they love. When the same rats are allowed to self-administer this brain stimulation they forget about the mazes, forget about their friends, and forget about pretty much everything else. They sit, pressing the lever a hundred times a minute for hours on end. They won't even take meal breaks, choosing to press the lever when they are famished and continuing until they die of starvation.

What would we do if we could stimulate a similarly intense pleasure center in our own brains? This question isn't hypothetical, of course, because we can. Consider the mother of all pleasure pathways: the orgasm. The positive sensation we feel is the release of chemicals that stimulate the same part of the brain that makes rats so happy. These "do-it-again" centers, when activated, associate pleasure with whatever behavior caused the brain stimulation.

Think of a do-it-again center as a square slot in your brain. Having sex is like discovering the proper square peg to fit the

slot. It makes us happy. The reward is an orgasm and this creates the desire to repeat the behavior, sex in this case. Having
discovered the square peg to the magic kingdom, we want to do
it again. (And again. And again.)

While we are busy enjoying our orgasms, our genes are laughing all the way to the Darwinian bank. From their perspective,
the result is (or was for our ancestors before birth control) having babies, which means the genes have successfully made it
into the next generation.

In their quest for immortality, our genes want us to perform a
variety of behaviors and accordingly have built many do-it-
again centers: imagine round and oval and star-shaped slots
permeating your brain. Genetically favorable behaviors have
been linked to these slots.

Eating a bit of strawberry shortcake, we are rewarded with
happiness as a round peg fills the round, calorie-seeking slot.
Achieving victory over a rival, we feel the euphoria that reliably
accompanies a star-shaped peg filling the star-shaped status
slot. In actuality, the variously shaped pegs are brain chemicals
that stimulate the do-it-again centers.

In creating such a pleasure system, our genes have built a reward system in which our pursuit of happiness accomplishes
their goals. No one has a baby because they want to replicate
genes, but by merely seeking pleasure and avoiding pain we unconsciously further our genes' goals. We needn't be aware of
genes at all; merely engaging in certain behaviors makes us feel
good and want to do them again.

Drugs hijack and short-circuit this evolutionary reward system.
Our ancestors got their chemical kicks the old-fashioned way:

they earned them through good behavior. With drugs, our pleasure centers can be stirred without the essential behaviors at all. How do drugs hijack the pain and pleasure routes in our brains?

We need to recognize that our only true erogenous zones are in our brains. In some completely paralyzed men, for instance, it is possible to stimulate the genitals to produce erections and even ejaculations. These patients, however, find no satisfaction because their brains never get the message. The same patients can, however, experience sensations like orgasms if the pleasure centers of their brains are stimulated. The trouble is, the brain has to be signaled about our behavior through the nervous system, and any signaling system can be manipulated.

Consider, for example, how predators lethally exploit the signaling system of the firefly. If you sit in a field on a summer night, you may be treated to a whirl of fireflies flashing in the dark. This dance is not for our pleasure; they are performing a mating ritual. It's pitch black in the field, and many different species are flying around. The flies need to find members of their own species in order to mate successfully, so they use a special Morse code signaling system that says, "Hey, I'm your type and I'm ready for action."

The fireflies don't actually see their potential lovers but instead communicate with belly lights. One species may beckon with two long flashes and a short while another may use four shorts followed by a long. When a sexually charged fly detects the right series of flashes, he or she swoops in, ready to begin a family.

Some of these flying Romeos and Juliets receive a rude shock. Arriving at the signaler, tiny loins aflame, they find jaws of death, not arms of love. Devious predators take advantage of the signaling system by producing the exact sequence of flashes sent by a willing mate. When a fly comes a-courtin' at the wrong home, it's dinnertime for the talented predator.

Our brain's signaling system can be similarly tricked — with disastrous consequences. When we do something good, our pleasure is caused by chemicals called neurotransmitters that stimulate our brains' do-it-again centers. Drugs — whether recreational or therapeutic, whether found in nature or made in the laboratory — mimic neurotransmitters. Just as firefly predators duplicate the flashes of a real mate, drugs "look" just like our natural chemical signals. Remember, caffeine works because it is so similar to adenosine.

When we take a pleasure-causing drug, our brain acts as if appropriately released neurotransmitters were flooding the system. The brain thinks we have done something great, such as finding food or warmth, when in fact we may be crouched over a filthy toilet with a hypodermic of heroin in our arm. Our pleasure centers know only that they are bathed in a precise set of chemical signals that induce bliss.

⌐

From alcohol to Prozac. Remember those rats that starved themselves to death as they pressed a lever to tickle their brains? They were stimulating one of the brain's "do-it-again" centers to release dopamine, one of the body's chief "happiness neurotransmitters." If we receive the dopamine message, the pleasure center in our brain makes us feel very, very good — so good, in fact, that we want to repeat our behavior.

As long as dopamine is bathing cells, those cells cause the intense pleasure we crave. Such pleasure is usually short-lived, however, because almost as soon as the message is sent, the dopamine is recycled back into the cell that released it.

When someone snorts a bit of cocaine, it heads straight for the brain's "do-it-again" centers. Once there, as with caffeine, everything hinges on a case of molecular mistaken identity. The cocaine fits snugly into the sites where dopamine is normally absorbed by the cells that originally released it. As long as these reuptake sites are blocked, the brain is bathed in higher levels of dopamine than usual; the cocaine user knows only one thing: let the good times roll.

Antidepressants work by an almost identical mechanism. In addition to dopamine, our body's other big happiness neurotransmitter is serotonin. The antidepressants Prozac and Zoloft, for instance, block serotonin from being recycled by the cells that released it. People are happier because of serotonin's prolonged stay in the synapse, lighting up our do-it-again centers like a pinball machine.

Some chemical messengers we ought to be especially thankful for are the endorphins, our body's natural painkillers. Produced by our brains, endorphins block pain messages arriving from throughout the body. Under a variety of situations of extreme stress — say we've just been seriously injured in a fight or we're in mile twelve of a half-marathon — our body responds by releasing endorphins. These chemicals also trigger the release of dopamine in the pleasure center.

The popular opiates morphine and heroin mimic endorphins, fitting snugly into their receptor sites. With a large enough dose, opiate users can give themselves an "endorphin rush" far more intense than anything possible with their own natural supply of these pleasure compounds.

One of our favorite drugs is tobacco's best friend, nicotine. Shortly after entering the bloodstream, nicotine begins mimicking one of the body's most common and important neurotransmitters, acetylcholine. Fooled by nicotine, acetylcholine receptors cause the release of adrenaline, other stimulating chemicals, and more of that pleasure-causing workhorse, dopamine. Nicotine causes rapid surges, then rapid depletions, of these chemicals, leaving the smoker happy for a short while but soon yearning for another cigarette.

Rats given daily infusions of nicotine bump up their acetylcholine receptors by 40% in less than a week. The human response is almost identical. It's odd that the body would build more of these receptors; regardless, their functioning diminishes with increased exposure to nicotine. After prolonged nicotine use, we're responding less and less despite inhaling more and more. However, with so many more receptors and with nicotine cleared from all the synapses as we sleep, the stage is set for maximal nicotine impact with the first cigarette of the morning.

So far, we've seen that drugs generally mimic chemicals used by our body during normal functioning. Their specific effects are quite predictable as long as we know the molecule that the drug mimics. Their work is like a surgical strike, altering our neurochemistry in a specific way.

But what happens if the drug is more of an everyman, looking enough like many different neurotransmitters to impersonate them all? Meet our friend, the cocktail. Alcohol is the great impersonator, fooling at least four different receptor molecules. In a quick survey of the functions of these victims, we can see exactly how alcohol works its magic.

1. *It slows us down, "relaxing" our neurons.* By blocking receptors for our brains' chief excitatory neurotransmitters, alcohol coats the brain in a bit of molasses, slowing reaction times and slurring speech. We could probably do without this effect.

2. *It gives us a pleasant buzz.* Acting like cocaine — but much weaker — alcohol blocks dopamine reuptake, increasing the concentration of the happy neurotransmitter in the key parts of our brains.

3. *It blocks pain.* By stimulating the release of endorphins, alcohol lets us sample the "runner's high" without even putting on our running shoes. Resembling morphine and heroin in this respect, but again at a greatly reduced magnitude, alcohol spurs our body to produce a little opiate-like high.

4. *Alcohol makes us happier, at least while it's in our system.* Like a "do-it-yourself Prozac kit," alcohol modifies and increases the efficiency of our serotonin receptors.

For all these reasons, many of us enjoy a glass of wine with dinner or the occasional cocktail after work. But what happens when one Cabernet becomes three, when one martini becomes too many to remember?

When the novelty fades. In American high schools, about half a million students — one third of them female — take muscle-building steroids. Consider what happens when one of them (let's call him Captain America) goes "on the juice," injecting himself with massive doses of testosterone. With all that extra steroid circulating in his veins, Captain America becomes bigger and stronger. As his muscles grow, however, the systems that regulate testosterone levels wonder, "Where did all this juice come from?"

Ever the adaptable machine, Captain America's body responds by cutting back its production of testosterone. The Captain continues injecting artificial testosterone, bulking up his physique even as his body scales back its natural production. Finally, he is confronted by a grim reality: his testosterone production sites, a.k.a. his testicles, have all but disappeared.

Here's the shrinking testicle lesson: our bodies don't like change. They work only within tightly set parameters. While a car can be ice cold or burning hot and still run beautifully, we die if our body's temperature changes a few degrees. We face similar rules when it comes to internal chemicals. Because we have so little latitude, we have evolved an array of systems that fight change. Go on a diet, for example, and we're foiled by a slowing metabolism.

An alcoholic woman recently checked herself into UCLA's hospital. Although her blood alcohol level was high enough to kill most people, she hadn't had a drink for almost three days. Her alcohol tolerance was so high that she needed enormous quantities to get a buzz, and three days later her blood was still about the consistency of a gin and tonic.

Like Captain America and hard-drinking alcoholics, regular users of many drugs develop tolerances. Many of us need a cup of coffee in the morning just to feel normal, and the average American consumes 225 milligrams of caffeine a day (the amount in five cans of Coca-Cola).

One study investigated caffeine tolerance by paying people to consume exactly 900 milligrams of caffeine a day for three weeks. Initially the caffeine users felt as wired as you'd imagine. Their high-flying ways did not last, though. Within three weeks, our caffeine freaks were indistinguishable from "clean" control counterparts. Whether rated for energy or alertness, tension or anxiety, caffeine exhibited no measurable influence. Complete tolerance had set in.

Tolerance is inevitable, but its costs and consequences vary by the type of drug. In another study, volunteers were injected with a uniform daily dose of heroin and monitored for their level of euphoria. (College students: pay attention to the fine print in campus newspaper ads.) Initially ecstatic, their bodies reacted by reducing the number of receptors that bind heroin. With fewer and fewer receptors, the heroin effects dropped almost to zero in just three weeks.

In the real world, heroin addicts do a little fighting back of their own. Over the course of an addiction, they may increase their dose ten-thousand-fold. If caffeine tolerance grew equivalently, we'd need to consume a bathtub full of coffee just to get out of bed. Fortunately for our bladders, caffeine tolerance rarely exceeds ten or fifteen times the original effective dose.

If we build up tolerances, why not just give up on the drugs? If only we could. The flip side of drug tolerance is unavoidable withdrawal pains. Our bodies will adjust to the absence of drugs, but repair takes time. When Captain America goes off the juice, his testicles need weeks to grow back to full size.

It's the same story for those using caffeine, nicotine, or alcohol. The costs range from headaches to the life-threatening *delirium tremens* of alcohol withdrawal. While we must all pay a withdrawal fee if we reduce our caffeine and other drug use, some of us are more easily hooked than others.

∽

One person's recreation, another's addiction. Isabella joins her friends in sipping wine during a dinner party. As the meal progresses, her companions become tipsy. Their conversations turn racy, their moods relaxed. They refill their glasses, reveling in a little buzz. Not so for Isabella. Before her first glass is empty, she experiences a "fast-flush" response: her face turns crimson, her heart begins to race, and her head starts to pound. Worse still, she soon feels the need to vomit.

How can people respond so differently to booze? Fast-flushers like Isabella have a genetic difference that causes the buildup of a poisonous chemical called acetaldehyde. When we splash a bit of alcohol down the hatch, our bodies do a little two-step dance in which they manhandle the alcohol molecules, converting them from their intoxicating form into innocuous atoms.

Isabella's body adeptly starts the normal breakdown of alcohol, but she was born with defective genetic instructions for making an enzyme that disposes of the poisonous compound. One by one, the alcohol molecules are processed, but without the correct machinery, poisonous acetaldehyde accumulates. Hence the "fast-flush" reaction.

Isabella's messed-up enzyme is called aldehyde dehydrogenase, and fully half of Asian people have the same genetic mistake. But hold on. Perhaps we ought to call this mistake a molecular godsend. In a study of thirteen hundred alcoholics in Japan, guess how many were fast-flushers? *Not one.* Although half the Japanese are fast-flushers, there was not a single one among these alcoholics. A minor change in their genetic code helps them resist the lure of alcohol.

So certain inborn genetic differences result in a decreased desire for drink. Is the opposite true as well? Do some individuals have genetic endowments that give them an unhealthy passion for the stuff? Animal studies suggest that this may be true.

Generally, mammals — from wild primates to household pets — avoid alcohol; given a choice, they choose water. Some scientists set out to breed rats with a taste for liquor. Each generation, only those animals with the least aversion for alcohol were selected as breeders. The rest, allowed no babies, remained heirless. The scientists soon had a rat population that loved to drink.

Interestingly, these alcohol-loving rats produced abnormally small amounts of the happy neurotransmitter serotonin in their brain. Their preference for alcohol may be an attempt to bump their serotonin levels back up to those maintained by normal rats. This finding opens a messy can of worms.

Do human drug addictions and dependencies reflect differences in our genes? Recent data suggest that they might. One group of scientists examining brains during autopsies found that the alcoholics had fewer dopamine receptors than the non-alcoholics.

Genes are implicated in other addictions, too. In one study of 283 individuals, a third of the people who smoked had an unusual copy of an important gene whereas almost none of the non-smokers carried it. This gene, labeled D2, enables the pleasure centers of our brains to light up when tickled by dopamine. Smokers with the unusual form of the D2 gene produce a third fewer dopamine receptors than normal.

Given dopamine's central role in orchestrating the pleasure centers of the brain, any alteration in this system wreaks havoc on the body's natural ability to regulate and achieve happiness. Many smokers can be viewed as medicating themselves in an attempt to stimulate their dopamine systems more aggressively. By smoking they can light up their pleasure centers to the level naturally enjoyed by the non-smokers.

As we saw earlier, other drugs, too — notably cocaine — prime the body's dopamine system. The same renegade D2 gene that predisposes people to smoking is also linked with other drug addictions and even to overeating.

In 1997 a man in North Carolina, Thomas Richard Jones, was tried for causing the death of two women in a car accident while he was under the influence of alcohol, painkillers, and antidepressants. Noting a long history of addiction, Jones's defense attorney pleaded that "the devil lurking in this alcohol and in these pills would not turn loose of him." One of the devil's aliases may be dopamine.

This is the danger in having a brain that uses chemical signals like dopamine and serotonin to regulate pleasure and happiness. Genetic glitches in the production of these chemicals may sentence some people to lives spent in search of a chemical high.

While genes have thus been shown to play a role in smoking, drinking, and the use of other drugs, we have clear evidence that genetic factors are not the whole story. Identical twins show similar — but not identical — propensities for drug use. If a person has a problem with alcohol, an identical twin is 25%–40% more likely than a fraternal twin to exhibit the same behavior. If genes were the whole story, identical twins would have identical behaviors.

These are the early days of understanding our brain's signaling systems. A complete understanding of addiction must incorporate both genes and the large set of non-genetic forces pushing people into or away from drug use.

⌐

Willpower to the rescue? "Just say no" to drugs is the simplest way to kick a habit. Unfortunately, this obvious and low-cost approach is also the route most likely to fail. For example, in any given year only one person quits smoking for every twenty who attempt to just say no. Raw willpower seems like a great solution right up until weakness strikes and we light up a cigarette or mix a margarita.

Alcoholics Anonymous and similar programs can be viewed as super-duper, augmented willpower. They provide members with a complex support system, but at their core, success relies on individual restraint. Even twelve-step willpower doesn't work very well. Critics say just 5% of AA members stay sober for a year. Supporters of AA dispute the 95% failure rate, but whatever the exact figures, willpower provides no magic solution.

Our willpower failures are demoralizing. Surely, we feel, if we get tougher, we can stay clean. Furthermore, we are often surrounded by people who do not understand addiction. Fourteen percent of Americans will, at some point, have a serious problem with alcohol. As stunning as this number is, it also means that 86% of Americans will never have such a dependency. This disapproving majority appears to say that clean living requires just a New Year's resolution and a bit of moral courage.

The inability to control drug passions lies not in personality defects, but in the strength of our urges. For some the desire is overwhelming. The golfer John Daly was willing to pay three million dollars for a drink. Thomas Covington, a crack addict who was arrested thirty-one times, said he used drugs in spite of imprisonment and fines because "once that compulsion is there, it doesn't matter what the penalty or the threat is."

Far from being an act of pure volition, addiction has powerful evolutionary and biological roots. Subtle differences in our brain wiring make us more or less susceptible to chemical manipulation. Although our weaknesses vary and most of us are spared the extremes of John Daly and Thomas Covington, every person has strong, instinctual cravings for destructive substances.

Because drugs hijack our genetic pleasure pathways, the substance abuse battle we fight is with ourselves. When our neurons experience the euphoria of a dopamine bath following the inhalation of a dopamine reuptake inhibitor like cocaine, our brain is in heaven. Never mind that we know we shouldn't be doing drugs or that part of us doesn't want to do drugs. It's like rewarding a puppy with petting and a big bone each time she urinates on the couch. Do you think she's going to learn to pee outdoors?

Quitting drugs can be similarly challenging. Like instructing ourselves to stop wanting food or love, our brains just can't take such a command seriously — to stop the behavior that generates our brain's highest reward. So we can't possibly be surprised that willpower alone is rarely sufficient; we shouldn't just try to "handle" an addiction that is taking over our lives.

For those who have never tried drugs, abstinence may indeed be the best strategy. This is particularly true for people who have a family history of addiction. It's easier never to start than to quit. But for those who are addicted, "just say no" just won't cut it. Fortunately for them, science is looking for ways to help.

Technocures. Thousands of years before the birth of Jesus, people who lived in Sumeria (part of modern Iraq) drank beer. They loved alcohol so much that the pictograph for beer is commonly found in ruins. Similarly, the Maya, native Central Americans, reportedly used the hallucinogenic compounds found on toads long before the arrival of the conquistadors. The Maya drug concoction was introduced into the colon as a "toad enema." Many other ancestral peoples also ingested drugs that occur naturally.

This casual drug use apparently caused few problems. These naturally occurring drugs, however, can be made much more potent by modern chemistry. Many people chew coca leaves, for instance, and get a caffeine-like buzz. While coca leaves contain less than 1% cocaine, the concentration and lure are magnified when it is refined to 60% or higher levels of purity. Crack cocaine reportedly produces a more intense sensation of pleasure than any natural act, including orgasm. No wonder people ruin their lives and betray their families to gain this high.

So modern technology takes a relatively innocuous product and purifies it into a destroyer. At least fourteen million chemicals exist; over the centuries, drug producers have honed in on the ten or twenty that just happen to tickle our neuro-fancy. It's no coincidence that recreational drugs are exactly those compounds that most effectively mess with the pleasure circuits in our brains. Like the ultimate double agent, however, technology may also be our most powerful ally in the drug battle.

Think back to *A Clockwork Orange*, directed by the late, great Stanley Kubrick. In the film, a futuristic society is filled with young thugs whose chief source of pleasure is mayhem and violence. Traditional efforts to restrain them result in large, expensive police forces and overflowing prisons. (Sound familiar?)

Rather than restraining the youths, the authorities alight on a "reform" strategy. The police reeducate the lead thug, Alex, by torturing him as he watches violence on film, and eventually he becomes physically ill at the mere thought of violence. The *Clockwork Orange* plan is repressive and ultimately fails. Nevertheless, it highlights an alternative strategy for drug control. Rather than restrain destructive passions, can we nip these powerful desires in the bud?

Recall those fast-flushers, who have trouble processing alcohol. While some people love nothing more than a good martini buzz, fast-flushers get sick from drinking alcohol. Consequently, they become alcoholics at much lower rates; they just aren't that interested in drinking. For fast-flushers, sobriety doesn't require any willpower. Can we harness this information to help others?

Imagine that you've got a magic pill that, once taken, makes a patient into a fast-flusher. Such a pill has been available for about fifty years. Called Antabuse, it deactivates the body's alcohol-processing machinery. With Antabuse, drinking causes poisons to build up and causes the nausea of the fast-flush response.

Antabuse seems perfectly designed to foil alcoholism. Most studies conclude, however, that it is of minimal help in treating alcoholism. How can that be? Take a look in the user's bedroom or garbage or toilet. Stories abound of alcoholics who flush their daily pill down the toilet or "cheek" it, only to dispose of it later. Users are adept at fooling their loved ones — one alcoholic's wife discovered a month's supply of Antabuse lined up above the kitchen doorframe.

Although Antabuse is on the right track, its effect disappears too quickly. An alcoholic discards a pill so he can get a pleasant buzz sometime within the next several days. To be effective, new anti-addiction chemicals must require only a moment of willpower while conferring longer lasting self-control. Imagine how much more effective such a remedy would be if it could be taken once a year or even given as a childhood vaccine.

There are a host of such passion-killing remedies under development. For example, a nicotine vaccine recently showed a lasting ability to reduce the pleasure from smoking. Another drug, BP 897, is a two-faced cocaine-fighting chemical. When someone is clean, it minimizes cocaine cravings and stimulates the dopamine system just a bit. As soon as a line of cocaine is snorted, BP 897 turns on the user, blocking the drug's effect.

These products and others being developed promise to dramatically augment the options. Increasingly the dilemmas will become like those of *A Clockwork Orange.* Can the government require people to take these drugs as punishment for certain transgressions? Should we "immunize" our children even if it deadens some of their passions? Creating nicotine vaccines and a long-lasting version of Antabuse could go a long way toward helping us deal with drugs. But there is another way that technology can reduce the costs of drug use.

The cigarette that we smoke contains pleasing, addictive nicotine. Prison is said to be unpleasant, not so much because of the loss of freedom but rather because of the company one must keep. Similarly, much of the damage from smoking comes not from the nicotine but from the other products in tobacco. Science now allows us to get our nicotine buzz through a patch or chewing gum, without any of the cancer-causing parts of tobacco.

By using nicotine patches or gum we are much more likely to quit smoking. As noted, only 5% of people who try to quit smoking on their own succeed. In one study of four thousand smokers, more than 40% remained smoke-free for a year by using a regimen that included nicotine supplements. These people aren't free of their nicotine addiction; technology allows them, however, to tickle that fancy with fewer side effects.

Methadone is another example of a drug designed to hijack the hijackers. Like heroin, methadone triggers our natural euphoria systems. In the United States, there are more than a hundred thousand former heroin users who now use methadone and can live relatively normal lives. Like those who wear a nicotine patch, methadone users have not shed their desires; they just satisfy their cravings with fewer consequences. A character in *Annie Hall* summarized this nicely: "I used to be a heroin addict, now I'm a methadone addict."

For the foreseeable future, we will live in a world where tobacco companies, spirit manufacturers, and drug cartels peddle all manner of destructive yet desirable chemicals. Although the current solutions are limited and imperfect, technology promises two paths to longer-term success. Increasingly, we will be able to dampen our passions or consume novel chemicals that let us have our pleasure cake and eat it, too.

Risk Thrill-seeking genes take us for a ride

Taking risks is costly but fun. Have you heard about the new state lottery game? It's called "Take a dollar and throw it in the trash." Actually, that's not quite fair. State lotteries return about 50% of the amount wagered, so the game is more aptly called "Take fifty cents and throw it in the trash." Given these odds, it may seem crazy that Americans wagered an average of $150 each on state lotteries last year, losing more than twenty billion dollars.

Once confined to sailors' quarters and back alleys, gambling has gone mainstream. Las Vegas and Atlantic City still draw millions but now compete with Native American casinos, riverboat gambling, ubiquitous state lotteries, and a nascent Internet gambling community. Soon we'll never be more than nanoseconds away from satisfying the urge to impoverish ourselves and our families.

Furthermore, two and a half million Americans suffer from severe gambling problems. And bets come disproportionately from those who can least afford the losses: lottery players with household incomes of less than $10,000 spend three times as much on tickets as those with incomes of more than $50,000. Cumulatively, Americans lose more than fifty billion dollars a year on legal gambling.

Why do we derive pleasure from making terrible bets? Are we simply the victims of advertising campaigns, greedy casino owners, and state governments looking for easy cash? Unfortunately, finding a scapegoat isn't so easy.

Looking across human cultures gives us a sense of what we're up against and shows us that gambling is universal. Indeed, casinos and other forms of gambling prosper from Vegas to Monte Carlo to Hong Kong. This taste for gaming extends even to non-industrialized societies. The Hadza, for example, are an African people who still live by hunting game and gathering plants. Despite never having seen a TV advertisement for the lotto, Hadza men spend so much time gambling that they are said to prefer games of chance to chances of game.

This universal love of gambling is just a small part of our general tendency to derive pleasure from taking risks. Anyone who has enjoyed driving a car a bit too fast knows the rush of a little danger. We watch movies about rebels without a cause, not about people buying insurance. Advertisements are filled with rock climbers and bungee jumpers, but rarely with favorable images of cautious people in their living rooms wearing helmets and safety goggles.

We are so fascinated by danger that we take risks even when we must pay a price. Why?

Part of the explanation is that we are terrible mathematicians. We can't seem to calculate odds correctly. Take the big drawing in the California state lottery, for example, where a bet consists of picking six numbers between 1 and 51. To win, these six numbers must match, in any order, the six randomly drawn numbers. Given these rules, what is the chance of winning? Write down an approximate answer.

Here's another brainteaser. Chinese families place a high value on sons, yet the Chinese government exerts extreme pressure to limit family size. Let's assume that the chance of having a girl is exactly 50%, but every couple stops having babies once they have a son. So half of the families have just a single boy, a quarter of the families have one boy and one girl, an eighth have one boy and two girls, etc. In this scenario, what percentage of Chinese babies will be male? (The answer is below.)

One more. Imagine that you are a doctor and one of your patients asks to take an HIV test. You assure her that the test is unnecessary as only one woman out of a thousand with her age and sexual history is infected. She insists, and sadly the test result indicates viral infection. If the HIV test is 95% accurate, what is the chance that your patient is actually sick?

Here are the answers. Let's start with the HIV test. When doctors and staff at Harvard Medical School were asked this question, the most common answer was a 95% chance that the patient was sick. They missed the mark by a mile; the correct answer is less than 2%. (We will explain later in the chapter.)

The Chinese population remains 50% girls even with the "stop at the first boy" rule in place. Finally, the chance of winning the California lottery is one in eighteen million: a person is nine times more likely to die by falling out of bed.

Don't feel bad if your answers were way off; so were ours, and that's exactly the point. Human statistical abilities are simply terrible for these sorts of questions. Our risk analysis troubles go on and on. We fear plane crashes more than car crashes although the risk of dying in an automobile is much, much higher. After a coin toss comes up heads five times in a row, we believe that tails is more likely on the next toss. And so on.

We are left with two puzzles. Why do people derive pleasure from taking risks? And why are we so bad at calculating the odds for those risks?

〜

Cautious Lassie? Animals often appear to be averse to risk. For example, when two four-hundred-pound red deer stags compete for a female, they rarely fight. First the two males stand next to each other and roar. If one's roar is more impressive, the other retreats with his proverbial tail between his legs. If the roars are equal, the males conduct a "parallel walk," where they check each other out while strutting their stuff. If one is significantly smaller, he withdraws.

Both roaring and the "rut strut" are risk-free mechanisms to determine which male would win in a fight. Larger, healthier animals can make more noise, and the inspection phase reveals an animal's size and musculature. Only when the two tests reveal an even match do competitions escalate to physical confrontation. And even then, fatalities are rare.

This avoidance of lethal combat in favor of simple sizing-up strategies is prevalent in many animals, from mammals to birds to insects, and suggests that animals shy away from risk. Upon closer inspection, however, we find that animals often *do* take life-threatening risks.

Some spider species' behavior resembles that of the roaring, parading stags. When two male spiders spot a female, the males assess each other and the smaller one leaves. In one devilishly cruel but clever experiment, a researcher allowed a really small male, let's call him Mini, to begin having sex with a female.

In these spiders, it isn't until several hours into the act that the male begins fertilizing the female. Exactly when Mini's sperm was beginning to work its genetic magic, the researcher introduced a much larger male, Hulkster.

What do you think happened? Mini is about to get a huge genetic payoff, so he's opposed to early withdrawal and ready to brawl. Hulkster notes the size difference and wades into battle, confident that Mini will flee. With two willing fighters, 90% of confrontations result in death or serious disfigurement. The smaller male loses 80% of these battles and almost always dies from his injuries, but 20% of the time the little guy wins and scores a rich genetic prize.

The message is that animals take risks when it pays. A little spider that slinks away will live, but he is unlikely to find another equally fertile female, so his risk-averse genes will die with him. Mothers in many species will risk death to protect their babies for the same genetic reason — they want to win the evolutionary competition. The winners pass their genes — and their instincts for risk — on to their offspring.

Human risk-taking is no different. We know, for example, that humans first arose in East Africa and then spread out to cover the rest of the world. Imagine two types of humans, those who cowered in their caves and those who explored new areas. While many of the risk-takers died, those who gambled and won populated the entire globe.

Across diverse cultures today there remain some clear, genetic benefits to taking risk. The Yanomamö, for example, are native South Americans who survive by hunting and small-scale farming. They are a fierce people; more than a quarter of the men die from violence. A man who has killed at least one other man is called an Unokai, and they are often murdered by relatives of their victims.

Why do Yanomamö men risk killing another? Those who do and survive end up with more wives and more babies. In one extensive, long-term study, 137 men were Unokais and 243 were not. The Unokais had, on average, 1.63 wives (polygamy is legal) and 4.91 children. The non-Unokais averaged only 0.63 wives and 1.59 children.

So we now see why people derive pleasure from taking risks. In their natural settings, humans and other animals take a risk when it's the smart thing to do. We are descended from the humans who left their caves, who took risks and won.

⌇

Built for risk. Our genes cajole us into taking risks by making danger exciting. We all receive a jolt when we ride a roller-coaster or a motorcycle. Risk triggers a biochemical reward system in which our brain produces dopamine, a chemical that makes us feel very good.

Rod, an acquaintance of Terry's, thrives on dangerous adventures around the world and also makes huge wagers. Hearing about some of Rod's risky exploits, Terry asked, "Does Rod like spicy food?" Absolutely. Rod not only loves a good jalapeño but also carries a bottle of hot sauce with him at all times.

Taking his love for spices to an extreme, Rod entered a chili-eating contest in which contestants paired off eating hotter and hotter peppers. In the finals, Rod won by eating a pepper so hot that half his face went numb (and stayed that way for nearly a week). His opponent gracefully backed out rather than trying to match the feat. What's the connection?

Risky behavior stimulates the dopamine reward systems. Some people are born with systems that muffle the buzz they get from taking risks. People born with these unusual dopamine receptors — and hence reduced stimulation of this pleasure pathway — will go to extreme lengths in pursuit of the dopamine high. They are risk freaks: bungee jumpers, racecar drivers, and explorers. Impulsive and extravagant, they are the highest of rollers in Vegas. Like Rod, they also tend to prefer spicier foods than others.

The press calls this a "novelty-seeking" gene. Recent evidence even shows a strong link between the prevalence of this single gene in a population and how far that group has migrated. Recall our observation that humans arose first in Africa, then migrated around the world. The longest migration of all was to South America. From Africa, these people went through Asia, across the land bridge to North America, and then all the way south.

Indigenous South Americans are the descendants of people who moved over and over again for thousands of years. More than two thirds of these people have the novelty-seeking gene — the highest prevalence of any group, and far higher than for modern Africans and Europeans, only a quarter of whom have the gene.

Other genetic differences influence our risky choices. For example, the less monoamine oxidase (a chemical regulator in the brain), the more likely a person is to crave excitement and take risks. Over time, we are likely to find more genes that cause such behavior, and we will also decipher the molecular methods used by these genes.

Some of us love a good bungee jump more than others. Even those of us with the standard dopamine receptors and ordinary levels of risk molecules are not immune to the rush of a risk, hence the broad appeal of amusement parks and casinos. Still searching for that scapegoat? Look within; our genes have made us risk junkies. Just as we are chemically seduced by the high of illegal drugs, we are drawn toward danger and the chemical cocktail it stirs.

Our genes even go a step further. They have built into our nature an unwarranted optimism, which in turn tricks us into overestimating our odds of winning. On the radio show *A Prairie Home Companion,* Garrison Keillor discusses the fictional town of Lake Wobegon, "where all the children are above average."

It is a mathematical fact that, as a group, we must be exactly average. When asked, however, we confidently state that we will live longer than others, get sick less frequently, and even pick stocks that will outperform the market. In one study, 94% of men ranked themselves in the top half of male athletic ability. Our overconfidence even allows people to believe they might win the lottery (and some, of course, do win). By creating such unrealistic beliefs, our genes goad us into taking greater risks than we might otherwise choose.

Part of the risk puzzle has been solved. Humans take risks because we are the great-great-grandchildren of humans who placed risky bets. Like other animals, our bodies and brains include systems that sometimes prod us down uncertain paths.

The second question remains, however. Why are people so bad at making risky decisions? Evolution should favor those who take the right risks, not just those who are reckless gamblers. If tiny spiders can figure out when it's a good time to roll the dice and take on Goliath-sized opponents, why can't we?

⁓

Animal geniuses. Some animals are surprisingly able statisticians. Consider that woodpeckers must decide which trees to peck on. Some trees are filled with tasty bugs while others are relatively barren. Humans attempting to solve this problem must resort to complicated mathematics. How do the woodpeckers make the right choice?

In a laboratory, woodpeckers were presented with two sorts of artificial trees. Both had twenty-four holes in them. In one group they were all empty. In the other, six of the twenty-four holes contained food. Like an oil wildcatter, a woodpecker should move on if the holes keep coming up empty.

But how many empties mean that the woodpecker should leave one tree and search for another? Leave too soon, and it deserts a rich area just because the first tests were unlucky. Leave too late, and the bird misses out on opportunities elsewhere.

Advanced mathematics yields the answer. In order to eat the most food, the birds should leave a tree after encountering six empty holes. What do the woodpeckers do? They look into 6.3 holes on average, almost perfect and far better than an untrained person would do with the same problem. When the experimenters increased or decreased the number of empty holes, the woodpeckers changed their sampling accordingly.

Woodpeckers aren't alone. Spiders, fish, stags, and many other animals solve problems that would require the mathematical prowess of an MIT Ph.D. How can their tiny brains do the job? Woodpeckers have faced similar problems for tens of thousands of years. The woodpeckers alive today descended from birds whose instincts solved the problem of finding food effectively. The saying "like a duck to water" reflects this very idea: animals instinctually solve problems common in their natural environment.

〜

Prehistoric fears. When it comes to predicting causes of death, our statistical reasoning would embarrass even a worm. For example, in an average year do more people in the United States die from natural disasters (tornado, flood, lightning) or diabetes? Americans believe that natural disasters are the bigger risk, yet in 1997, 62,636 Americans died from diabetes and just 227 from tornados, floods, and lightning.

Similarly, women fear death due to pregnancy, even though almost no one dies from birth-related complications. In 1997, 291 U.S. deaths were attributed to pregnancy (including those suffered by mothers during abortions) while 159,791 people died from stroke.

Beyond pregnancy, our fears and statistical estimates of death rates are wildly out of whack with reality. We most overestimate the risk of death from accidents, homicide, and venomous bites, and we most underestimate deaths from a wide variety of diseases and vaccinations. Are people statistical birdbrains? To make good risky decisions, it would be helpful to have accurate estimates of risk. But wait. All of the figures cited above are from modern America.

What risks would our ancestors have faced, before the advent of modern medicine and easy living? While we can't know for sure, we can examine what kills modern peoples who still live in non-industrialized conditions.

A long-term study of the Ache, a group of South American foragers, revealed that among 87 documented men's deaths, 12 were from snakebites, 7 from jaguar attacks, 2 from lightning, and 6 from club fights. Thirty-nine women's deaths included 3 from snakebites, 1 from a jaguar attack, and 1 from lightning. Three women died giving birth, almost 10% of the women's deaths.

Over the course of her life, an American woman today stands a 1 out of 3,700 chance that she will die from pregnancy-related complications. In Africa, more pregnancies and a lower standard of medical care combine to cause 1 of every 16 women to die from pregnancy. Ancestral humans presumably had death rates similar to those in modern Africa and among the Ache. In other words, pregnancy and its complications were one of the leading causes of death for ancestral women.

Before modern medicine, humans did not die quietly of strokes while resting in nursing homes. Furthermore, there was no treatment for most diseases. If a person fears a homicidal killer, she or he can take steps to save themselves. Fearing cancer in a world without hospitals has no similar function.

Our fears are indeed rational, not for a modern world, but for that of our ancestors. They were bitten by snakes, killed by animals (including humans), killed in accidents, and died during childbirth. We have inherited a set of instinctual fears appropriate to their world. As we continue our investigation, we will see that many judgment errors in human risk-taking stem from a common source: we live in a very different world from that of our ancestors.

Case in point: Publisher's Clearinghouse struggled for many years until it decided that instead of rewarding many entrants with medium-sized prizes, it would offer huge prizes with tiny odds. As one executive recounts, "People don't care about the odds, only the prizes." Although the value of the prizes increased, the chance of winning decreased even more. Amazingly, people flocked to the games. Why? It turns out that we are bad at estimating extremely unlikely events.

No one knows why we are so easily fooled in these situations. One possibility is that our ancestors evolved in a world with very few people. Our hunter-gatherer groups had about a hundred people and, not so long ago on a genetic timescale, there were only eighteen thousand humans in the entire world.

When we see lottery winners on TV, we are unable to wrap our brains around the enormous size of modern populations. They're unprecedented in our genes' experiences. Our instincts tell us that our chances must be pretty good, so perhaps we should emulate the winners. The reality is that we are much much less likely to win than we think we are. In the words of Buddy Roogow, the commissioner of the Maryland lottery, "It could be you . . . but it probably won't."

Let's return to the puzzle of the HIV test. We started by assuming that in each group of a thousand people, one person is sick and the rest are healthy. Now run the test on everyone. How many get a test result that says they are sick? Let's count. First, the one person who actually is sick will test positive. But among the healthy people, the 95% accuracy rate means that fifty (5%) will have test results that falsely report they are sick. So fifty-one people get bad news, but only one (2%) of them is actually sick.

In test after test, people botch problems involving percentages. Most people also get the Chinese baby problem wrong. Half of the families have a single boy, a quarter of the families have a boy and a girl, and smaller percentages have one boy and more girls. Seems like a lot of boys. In truth, since every birth has a 50% chance of being a girl, it doesn't matter how those births are distributed among families.

If our brilliant woodpeckers or daring spiders took standard IQ tests, they would score a big fat zero. Animals look really smart when they solve problems that have been important to them and their ancestors throughout their evolutionary history. It's easy to make them look foolish by putting them in new situations, the proverbial "fish out of water."

The same is true for humans. An important difference, however, is that while many animals still live in their ancestral environments, humans do not. So animals usually look pretty smart, while people often look silly.

⌐

Risky business. Like most people, Jay likes an occasional wager, and with his wife, Lisa, he hosts a periodic poker game. The game is most fun when big pots are on the line with numerous people upping the ante. Jay has discovered that the surest way to create these high-stakes situations is to designate a large number of wild cards.

People bet more with all the wild cards because they fail to recalibrate the increased likelihood of a strong hand. They are, for example, convinced that their full house is invincible until confronted with four of a kind that includes two wild cards. With several players confident of victory, betting is more likely to escalate and pots grow large.

Lottery officials have figured out the equivalent of wild cards and have created games that entice our gambler within while fooling our more calculating side. In fact, most of us have no idea what odds we face. As noted, winning the big drawing in California requires matching six numbers between 1 and 51. Why these rules? Precisely to hide the terrible odds. On mathematical problems of exactly this sort, people overestimate the odds of winning by more than one thousand percent. That's why lotteries use them.

Let's face it, if a friend said, "I'm thinking of a number between one and eighteen million. See if you can guess it," you probably wouldn't have much hope of winning nor would you wager the family's grocery money.

Businesses take advantage of our instincts in other ways. In one experiment, researchers ran a lottery with a twist. Half the people were allowed to pick their entries. The others were assigned entries at random. Just before the drawing, the researchers offered to buy the tickets back from the subjects.

What did they find? People who had been assigned tickets were willing to sell them for an average of just under two bucks, while those who picked their entry demanded more than eight dollars. This enormous difference seems silly. The lottery was run purely by chance, so every ticket, chosen or assigned, had the same value.

A related study had people play a game of chance against an opponent. The game was simply to have each person draw a playing card, with the higher card winning. Half the bettors played against a well-dressed opponent who acted in a confident manner. The other half's opponents were instructed to act in a bumbling manner and wear clothes that did not fit.

What is the chance of winning this game against the cool, powerful opponent? Exactly half, and exactly the same as the odds of winning against the bumbling fool. Remember, this is a game of total chance. In the experiment, however, bettors wagered 47% more when they faced meek, poorly dressed opponents.

Once again we have evidence of human irrationality. We dramatically alter our behavior based on irrelevant factors. In a game of pure chance, it doesn't matter whether we are playing against Albert Einstein or Forrest Gump.

Now put on your anthropologist's hat for a moment. In most real situations, are you more likely to defeat a competent whiz or a bumbling fool? Obviously, we almost always need to be aware of the competition. Would you bet on yourself in a tennis match against Venus Williams? Our behavior in games of pure chance looks silly, but as with our other foibles, our instincts work well in social interactions — more natural environments. Our genes tell us to assess our opponent.

Predictably, casinos and other purveyors of gambles take advantage of every one of our wagering instincts. We go to the craps table and place bigger bets on people who look competent; many lotteries let us choose our favorite numbers. If casinos were smart, they would make all blackjack dealers wear silly clothes. Oh, they do.

⏳

Can we have our thrills without the spills? *Saturday Night Live* has a tradition of showing fake advertisements for silly businesses. One was for the First Citiwide Change Bank, a financial institution that only makes change. Their motto: "You give us a ten dollar bill and we give you two fives. You give us a quarter, we give you two dimes and a nickel. How do we make money? Volume."

Regardless of how much volume the bank does, clearly the Change Bank can't make money if every transaction is a wash. Similarly, firms that buy and sell risk only make money by taking our cash. They can't turn a profit buying and selling risk at cost. For example, every single insurance policy costs more to buy than the average payout to a policyholder. That's the only way insurance firms can survive.

Just because insurance is a money-losing venture for customers, does that mean we should not have it? Of course not. When we buy any product — whether it's an automobile from Ford or a pizza from Wolfgang Puck — the seller prices it above cost. And we ought to buy insurance in many cases. Unfortunately, we are often fooled by products that have been artfully designed to exploit our risky instincts.

To make the right risky moves, the first step is to analyze the situation clearly. When a salesperson offers us home insurance or a warranty on a refrigerator, we need to recognize that we are being offered a deal that, on average, will cost us money. As with any product, we should buy these risk-reducing products only after making sure that we understand them and that they are worth more to us than the price.

Similarly, every bet offered by a state lottery or a casino is a money-losing offer. Many of us find these offers worth the cost, but as with insurance, we should know the score before laying down our cash.

For example, each and every dollar bet on a standard roulette wheel returns ninety-five cents. There are no better or worse wagers in roulette; you can think of each play as the purchase of a dollar's worth of thrill for five cents. In contrast, the best bets in craps can cost as little as half a cent per dollar bet, while the worst craps bets cost almost ten cents per dollar. So feel free to play roulette if the thrill is worth five cents per dollar wagered.

Don't play craps, though, unless you know the good bets and the bad ones.

The point here is not that we all need to learn the details of casino games (although that can be fun). It's that we needn't be suckers in games of risk. To triumph in these uncertain arenas, we must rely on mathematical analysis. While most of us can't do the required calculations, plenty of books and online resources can help us. We must use these resources — not trust our instincts.

Here's a little secret. Of all the problems we discuss in *Mean Genes*, love of risk is Terry's primary personal weakness. This was revealed early in Terry's life when his family played a game called Stocks & Bonds. Just like the actual stock market, players invest their paper money in various companies. A trading day is simulated by pulling cards from a deck that lists the change in price for each stock.

The game has some conservative stocks that move by a quarter point a day or less. There is also an oil company stock, Striker Drilling, that more closely resembles an Internet company. Some days Striker Drilling's stock price went up by twenty dollars, other times down by seventeen dollars As a kid, Terry invested exclusively in Striker Drilling, reveling in the ups and downs (and eating spicy foods while playing).

In 1998 those same risk-loving attitudes had Terry knee-deep in day trading, making more than two thousand trades, buying and selling a quarter of a billion dollars worth of securities. Although it was profitable, he decided all this trading activity was making him unhappy and boring. What to do? The allure of trading was almost overwhelming, however, and he broke numerous promises to himself about quitting.

The solution was twofold. First, Terry moved his accounts from a low-cost Internet broker to a traditional broker with higher fees and a slower response. Previously, when the urge to trade struck, Terry was just clicks away from a trade. With the new broker, it requires a phone call, two minutes of banter about golf scores, and a hefty commission.

The second step was to stop watching the stock prices moment by moment. With an Internet connection, anyone can monitor their investments just like a stockbroker. This was an incredible time sink for Terry. He began to force himself offline by giving his Internet access cable to friends for a day at a time. He even took to mailing it to himself when he felt overpowered by the urge. Terry has now reduced his trading toward zero — and consequently has room for other things in his life.

When passion strikes, it is usually too late to control ourselves. Willpower all too frequently lapses in the face of strong desire. The key to control is taking steps in advance that limit the gambler within. With respect to gambling, this insight translates into never, ever taking a credit card into a casino. Leave it locked up in the hotel safe, or better yet, at home.

Nowadays casinos will extend credit with a driver's license or less. Some ATM machines in casinos dispense cash but don't take deposits. To win, we need to decide in advance the maximum we can lose, then ensure we absolutely cannot exceed that limit.

When eating out, it is fun to play a game called credit card roulette. Everyone at the table puts a credit card into a hat or dinner napkin. The waiter is then asked to pick one card, and the person whose card is selected pays for the entire dinner. This is gambling, but unlike lotteries and casino games, it costs nothing. If played repeatedly, all it does is switch around who picks up the check. And although it's free, it's surprisingly exciting.

There are other ways to get our thrills at a low cost or even to be paid. All manner of inventions exist that make us think we are about to die when we are actually safe. Some roller-coasters now travel over 80 miles an hour and include plunges engineered to rev up our instincts. After a moment of terror, we hop off and have a candy bar. Similarly, horror movies, parachuting, bungee jumping, and a variety of video games all mix a chemical cocktail designed to give us a buzz with no hangover.

Historically, the mother of all free lunches, in terms of risk, has been the stock market. Buying U.S. stocks has proved to be a wonderful investment with thrilling ups and downs. Since the early nineteenth century, long-term investors who bought U.S. stocks made far more than those who owned bonds or gold. In the financial arena, we can take risks and get paid for them.

Investing, however, is an area where our instincts often get us in trouble. Remember that humans are overconfident in many areas, including finance. Money managers go through years of training, yet most of them fail to beat the proverbial dart. The story is far worse for individual investors. Studies show that the more actively we trade, the worse we do. So stocks can be both risky and profitable, but we have to shackle the overconfident day trader who lurks within each of us.

⌐

Go out on a limb. That's where the fruit is. Some years ago, John O'Connor was feeling bold while on a date with a woman named Sandra. Rather than simply ask her out for a second date, he went gung-ho and asked her for more than a dozen dates. What are they doing these days? Still happily married. (She is now U.S. Supreme Court Justice Sandra Day O'Connor.)

We've been looking at situations in which people like risk too much. In many cases, however, we have exactly the opposite problem: we are too timid. The social arena is one important area in which we ought to take more risks.

For ancestral humans, social failures were presumably much more costly than they are for us. Remember that, until recently, people lived with the same group from early adulthood until death. In such an environment, people were doomed to hear about their social mistakes for years as the same group gathered around the campfire to joke about Johnny the Overeager.

Far beyond a never-ending series of jokes, social mistakes could have had fatal consequences for ancestral humans. Single humans did not fare well in the dangerous world of our ancestors. Offend the wrong group of people, and a social risk could rapidly turn into a very bad day. Among the Yanomamö, individuals thrown out of their villages could sometimes join neighboring groups, but they also risked being killed.

So our ancestors were right to be timid in a variety of situations. In contrast, imagine what would have happened if John O'Connor had been turned down. No big deal, there are lots of people in the world, and he would have moved on. While we joke that everyone is connected by six degrees of separation, our ancestors had no degrees of separation.

A similar change has occurred in many other domains. Our ancestors got their thrills the old-fashioned way: they took risks. Consider the human known as the Iceman. About five thousand years ago, he set out on a European adventure and ended up frozen in a glacier. Completely preserved, he can now be visited in a museum.

For the Iceman, a career mistake had fatal consequences. If we make a risky move — say, by taking a job with some poorly financed Internet company — and it fails, we will not be killed. More likely than not, we will find a new job with a higher salary. If the failure is spectacular enough, we can write a series of books, like Donald Trump. If we just rely on our instincts, we are likely to be too timid in our professional lives.

Sir Edmund Hillary, who led the first group to reach the peak of Mount Everest, said, "I get frightened to death on many, many occasions, but fear can also be a stimulating factor . . . you can often extend yourself far more than you ever believed possible." The good news is that we don't need to climb mountains to enjoy the intense thrill of risk.

Greed Running fast on the happiness treadmill

Happy (pay) days. "If you don't think money can buy happiness, you don't have enough money," reads a help-wanted ad. Extra money, particularly when it comes as a surprise, is certain to brighten anyone's day. It seems obvious that with enough money, we could live happily ever after. Indeed, when Americans were asked to name the single change that would most improve their life, the most common answer was "more money."

In search of both money and career advancement, Americans are working harder than ever. And while the rich used to be famous for their idling, these days even the prosperous among us are spending more time at the office and taking shorter vacations. What has all of this hard work brought us? The short answer: mirthless materialism.

The average income in the United States (adjusted for inflation) has risen more than 40% since 1972. Every year, researchers have asked, "How happy are you with your life?" In spite of having more money, safer cars, and homes that have doubled in size, our answers reveal no change in satisfaction over this period. Similarly, the average person in Japan has become more than three times richer since 1958, and the Japanese too report no increase in happiness. So we are much richer, yet we are no happier.

The conclusion appears obvious but also puzzling. Deep, long-term happiness does not come from material circumstances. Although *acquiring* money, TVs, and cars makes us happy, *having* them does not.

In addition to cash, many other factors have surprisingly little effect on happiness. For example, Midwesterners overwhelmingly cite cold winters as a source of displeasure and imagine that Californians are happier. While Californians do love their sunny climate, however, they report themselves no happier than Midwesterners.

Skeptical? Perhaps some people are lying, saying they are happy when they are actually miserable or vice versa. (Maybe the Californians want to ensure that others don't move there.) Should we trust these self-reported levels of happiness? Because happiness is difficult to judge from the outside, there are few facts available beyond simply asking people how they feel. One of these admittedly imperfect measures of happiness — or, more accurately, unhappiness — is suicide.

If money made us happy, people in poor countries might be expected to kill themselves more than people in rich countries. They don't. Take Japan, for example, one of the richest countries in the world, where gleaming bullet trains whisk people to high-tech jobs. In 1998 Japan had one of the highest suicide rate in the world, slightly behind that of Finland, another rich country.

In the affluent United States, more than thirty thousand people killed themselves last year, and five hundred thousand more were treated at hospitals for suicide attempts. More young people in the United States die of suicide than from AIDS, cancer, and heart disease combined.

A materialistic view of happiness might also predict that in any country, the poor kill themselves more often. Wrong again. Suicide is the third leading cause of death for American teens, but among the richer kids who go to college, suicide is the number two killer. Similarly, African-Americans, who are poorer overall, commit suicide at significantly lower rates than other Americans.

Suicide is only the tip of the depression iceberg. Twenty-five million Americans suffer a major depression every year. With tens of millions more suffering severe bouts of unhappiness, we are justly labeled the Prozac nation. Rich people and rich countries have never had more material abundance. Yet, more of us than ever before are depressed and suicidal.

⌒

Built to want more. Who can we blame for our greed and unhappiness? Advertisements surely play a role in stoking our desires, but poor people around the world join American yuppies in striving for material goals. When technologically simple societies are exposed to Western goods, the people immediately want refrigerators, antibiotics, and Michael Jordan T-shirts, sensing that smiles are just an acquisition away. The lure of filthy lucre is strong and natural.

The Yanomamö, native South Americans, live in the tropical forest near the border of Venezuela and Brazil. In 1964, when the anthropologist Napoleon Chagnon went to live with them, the tribe had no TVs or other media to influence their desires. Their way of life — with little technology or industry — was like a window into our evolutionary past.

With only the crudest tools and weapons, the Yanomamö obtained a wide variety of foods. They harvested honey (which they love), tended gardens of plantains (sort of a cross between a banana and a potato), and hunted wild pigs, monkeys, birds, and even a few snakes. They also ate palm grubs, squirming, maggot-like morsels as large as a mouse (and, when sautéed, reportedly taste like bacon).

These people lived without a single TV commercial and not a single corporation sold them products, yet they shared our thirst for worldly goods. Soon after he arrived among the Yanomamö, Chagnon was robbed of nearly every item he had brought — clothes, tools, drugs, and food.

In fact, over the course of his more than five years among the Yanomamö, Chagnon lamented that he was constantly badgered for everything from matches to flashlights and axes. He became "bitterly distressed" that even his friends among the Yanomamö wanted nothing more than to gain access to his locked hut so they could steal.

Chagnon's experience mirrors that of other anthropologists. One of the most common stories from those living in non-industrialized cultures is the "knock" on the tent with a request for food, water, drugs, or weapons. The drive to acquire material possessions is a human universal. Aggressive ad campaigns may stimulate and aggravate the acquisitive beast, but that monster prowls within us all.

We conclude that permanent changes in life circumstances do not produce permanent changes in happiness. Although the data we just read are powerful, perhaps even more striking support comes from a study in which researchers interviewed people soon after they had experienced life-changing events. The lucky ones had won fat lottery prizes; the unlucky ones had suffered accidents that left them unable to walk (or worse) for the rest of their lives.

At the time of these events, as expected, the winners became ecstatic and the victims despaired. Over time, however, both groups moved back toward the average level of happiness reported by people who had neither won a lottery nor been in an accident. Before even a year had passed after receiving their windfall, the lottery winners reported an average satisfaction level no higher than that of the general population.

Participants in MTV's *The Real World* go through a similar experience. In a typical series, seven young people get to live for free in a multimillion-dollar house with lavish furnishings. They are initially blown away and thrilled with their new digs, but soon fall into angst and unhappiness.

So people quickly adjust to an improvement in their life. What about those people in the study who had suffered accidents and become paralyzed? Within a year of their accident, the victims reported an average happiness level of 3.0 on a 5-point scale, lower than the overall average of 4 but far above despair.

These results have become famous because the changes in happiness were less severe, and went away faster, than most of us would expect. Christopher Reeve's story describes the emotional path typically taken by people who experience tragedies like the accident victims in this study.

In the spring of 1995, Christopher Reeve became a quadriplegic after he was thrown from his horse during a competition. As a famous actor who was literally Superman, Reeve saw his privileged life among the stars reduced to wheelchairs, physical therapy, and sponge baths. As he recounts in his autobiography, he felt that he had ruined his life and wondered, "Why not die and save everyone a lot of trouble?"

In a few years, Reeve returned to the public arena, becoming an active crusader to increase funding for spinal cord research. He spoke at the 1996 Academy Awards and starred, with Daryl Hannah, in a 1998 remake of *Rear Window*. He remains confined to a wheelchair, with no ability to control his body below the neck, yet his "optimism remains intact" and, he confidently states, "When I look to the future, I see more possibilities than limitations."

Reeve's story parallels those of many people who survive tragedies. There are dark days after the incident, but hope reappears. This sequence of emotions is so common that it is the rule rather than the exception. Still, people do not predict they will get beyond their negative moods. Indeed, Reeve recalls being told that he would recover emotionally and he says simply, "I couldn't believe it."

So if you want to predict how happy someone is today, don't ask about their career, their income, their love life, or even if they can walk. Surprisingly, the most useful piece of data is simply a description of how happy he or she was at age twenty (or even age six).

Mood is substantially a function of personality; some people are just blessed with a positive outlook. Seemingly important information such as age, sex, race, or financial status provides almost no predictive improvement over simply assuming that there are happy people and there are unhappy people.

Knowing this, consider again how you would feel if a juicy, tax-free million fell into your lap? Would the ecstasy completely fade even as you tanned on some tropical beach counting stacks of crisp bills? In fact, you would be no happier a year after winning a million dollars than you are today. Impossible as it seems, our happiness, though influenced by ephemeral events, is not controlled by them.

This contrast between the powerful short-term effect of life changes and the minimal long-term effect is one of the central paradoxes of being human. To unravel this mystery, we must learn why our genes benefit from building people who have trouble finding happiness yet remain confident that it is within reach.

⌒

Chase the money rabbit. Owners of dog racing facilities have learned how to create exciting contests by using an artificial rabbit. The dogs think they'll soon be feasting on rabbit flesh, but they'll never catch their prey. To entertain the customers, the racetrack keeps the rabbit just ahead of the dogs.

Happiness is a tool that our genes use to induce us toward behaviors that benefit them. The rabbit moves to further the interests of the racetrack owner, not the dog. Similarly, we strive towards elusive goals, not for our own happiness, but to further the interests of our genes.

While we will never be finished, we are built to feel that permanent satisfaction is possible if we could just get a little bit ahead. Maybe catch a lucky break at work or in the lottery. Once the current crises pass, all will be better and our problems will dissipate. Because our dreams are always just a step ahead, we work inexorably to better our situation.

One consequence of this relentless system is that we adjust quickly to good fortune. In 1976, for example, Elvis Presley earned several million dollars. Unfortunately, he lapsed further into debt as his spending went up faster than his income. That year, he spent at a prodigious rate, including $35,000 on a single meal of peanut butter and banana sandwiches. (Elvis took some friends from Memphis to Denver and back on his private jet, the *Lisa Marie,* for the meal.)

Elvis was not alone in his ability to increase spending rapidly in good times. While we hope to live comfortably within our means, our instincts propel us to change our behavior so we are always living on the edge. We share this attribute with many other animals. Among opossums, for example, those with the biggest families spend the most time scurrying around to feed and protect them.

Why don't these furry parents cut back, have fewer babies, and enjoy more free time? Well, natural selection is a relentless taskmaster and favors those with large families, even if the animals have to run ragged tending their bulging brood.

One study looked at the effects of newfound riches on opossum family size. Researchers randomly selected animals to be the equivalent of lottery winners, giving the lucky ones huge amounts of good food. With this windfall, the recipients could have maintained their existing family size and gained leisure time. What happened? Like Elvis, our well-fed opossums adjusted immediately: they "invested" their extra resources in making bigger babies and soon were just as busy as before.

Surely there must be some limit. As we get richer and richer as a society and individuals, won't we reach a point of bliss? If each of us pays off our mortgage, acquires our dream car, and can afford good health care for our families, won't our greedy drives be satisfied? Sadly, our happiness racetrack has no speed limit and no exit.

Consider the shifting goals of Jim Clark, who became one of the richest men in the world by starting Silicon Graphics, Netscape, and several other companies. Before he was bitten by the business bug, Clark was a Stanford professor earning low pay and dreaming of striking it rich.

He told his friends that if somehow he could make a hundred million dollars he would be eternally satisfied. When he passed that lofty goal, he reset his sights on a billion. Now, with billions in his pocket, he is working just as hard as ever and hopes to seize the title of world's richest human from Bill Gates. Having founded three different companies, each worth more than a billion dollars, Clark still toils.

Lottery winners, accident victims, Jim Clark, and even Elvis's spirit chase elusive goals and run from fleeting fears. We all do. We've been built in such a way that the satisfaction game cannot be won by accomplishing goals nor lost by any setback. Allowing us to rest on our laurels or weep over spilt milk would be a genetic mistake. Our genes don't care about our past achievements, only about continually positioning our emotional rabbits just far enough ahead to keep us panting and working to accomplish their goals.

~

Catching elusive goals. Imagine that you work in sales for a demanding boss, who sets challenging goals but offers exorbitant rewards. For example, if you sell a thousand encyclopedias this month, the company will pay for a luxurious vacation. You work with the single-minded purpose of tasting that fine life. In the second week, you convince a university to buy an encyclopedia for every student, and you've hit your quota. Fat city, here we come!

You rush into the boss's office wearing a Hawaiian shirt and carrying your golf clubs. She says, "Wow, that was great, but there will be no vacation. Tomorrow, we start toward a new quota of eleven hundred. If you make this higher quota, we'll give you a fancy mansion."

In addition to being illegal, such an incentive system would not produce sales. Once deceived, only a fool would earnestly work toward the new quota. We are wise to people who renege on their promises. Unfortunately, we are less savvy when it comes to our own internal promises.

In *Peanuts,* Charlie Brown gets fooled by Lucy over and over again. She puts a football down and exhorts Charlie to kick with all his might. He reminds Lucy that in previous attempts, she has tricked him by moving the ball. Assured that this time she is being honest, Charlie runs full tilt and makes a vigorous kick. Lucy removes the ball at the last second and Charlie hits the ground with a thud.

In real life, Charlie would soon catch on to the duplicity, just like our encyclopedia salesperson. But the fact that the game is played over and over illuminates something deep in human nature. In our pursuit of happiness, we act very much like Charlie Brown, repeatedly running toward moving targets. We constantly think, "If I can just get through this week, everything will be okay from now on." Or, "Once I pay these credit cards off, I'll never get into this situation again."

Lurking inside our hopes are genes that want us to work hard all the time. They prosper most when we run full tilt. Once we approach the point of promised bliss, the emotional football is moved again. In this manner, we are motivated to do our best at every minute.

This biological drive also explains why we recover from catastrophes. Our genes help us avoid accidents by building us to fear certain situations and using pain to teach us to avoid repeating damaging behaviors. When tragedy strikes, however, hardness becomes compassion. Regardless of how devastating our calamity or how stupid our behavior, they forgive us.

Our instinctual systems are thus both the toughest and the nicest of bosses. They constantly ride us for more, not caring what we accomplished yesterday but asking for maximal effort today.

Our emotions are thus designed to be less permanent than they feel. For example, women report that they have trouble remembering the pain of giving birth. The evolutionary advantages of this convenient amnesia are obvious, and all of us who are not the first-born in our families should be thankful. For similar reasons, we cannot recognize the changing nature of our goals. The genetic boss makes us forget that last week's promises were not kept. Now that we are on to this game, how can we capitalize?

We can begin by trying to take our goals less seriously. We shouldn't buy things in the hope that they will make us happier. At the time of purchase, we trade off the joys of a faster computer or a bigger house with the costs of payment. The joy will fade — more quickly than we are built to anticipate — but the bills will remain. We need to learn that long after our new computer stops seeming fast, we'll still be paying it off.

A silver lining is that pain goes away faster, and hurts less, than we expect. We overestimate how depressed we'll feel when bad outcomes occur. Any sports fan knows that the hurt of losing fades and excitement rebuilds for the next game or the next season. More seriously, patients awaiting the results of HIV tests expect to be devastated if they learn they are infected, but afterward are much less depressed than they predicted.

Because we recover faster than we expect, we should take more chances. Many scary choices have limited downsides (frequently just humiliation). Social failures such as asking someone out or trying a new hair style can be emotionally crushing. Similarly, we may not change to our dream career because we fear a temporary downward move in responsibility, prestige, and salary. Learning that the pain will go away sooner than we predict, however, can help us become more courageous.

We should avoid making big decisions soon after dramatic life changes. Half of prison suicides take place on the first day of imprisonment. While we are mired in depression or bursting with joy, we have trouble believing that these strong feelings will dissipate.

We need to restrain our actions and take concrete steps to prevent impetuous decisions. We should not kill ourselves after a car wreck or give away our millions just after we earn them. Wait six months. If we get an unexpected financial windfall, we ought to lock it up immediately in a savings account that cannot be touched for six months.

We can also use our knowledge to predict the behavior of others. When Elizabeth Taylor divorced Richard Burton for the first time, was he surprised? What about when she divorced him after their second marriage? The best way to predict someone's future behavior is to examine their past. Burton was Taylor's fifth and sixth husband on the way to a total of eight. He should have expected their marriages to end just as her previous ones had.

People change far less than they think. When it comes to happiness, the most common story is that happy children grow up to be happy adolescents who then become happy adults. The best way to ensure that we'll be surrounded by upbeat people in the future is to be friends with happy people now.

Finally, we should start our own changes today. We are built to feel that tomorrow will be different from today. Unless we take steps to change, our state of affairs today is a much clearer predictor of the future than we imagine. We should try to live today as though it were every day. It is.

Too much is never enough. We become ecstatic when extra money drops in our lap, yet cash has almost no long-term role in creating happiness. This seems impossible; how can every extra dollar make us happier, but thousands of dollars have no long-term effect? It's as though we are on a treadmill. We march steadily forward but the "ground" moves backward, so that after considerable effort we have not moved forward even one inch.

We gain insight into the origin of this treadmill by asking, "How much is enough?" When it comes from our genes, the answer is "as much as possible." Evolution is a competitive game in which victory comes not from achieving some fixed number of points but by simply outscoring the opposition. We are descended from the humans who had the most children, not from those with "enough" children.

Imagine two kinds of people: the "reasonable" and the "greedy." Reasonable people are satisfied after amassing some wealth and then spend the rest of the day playing the harp and tickling their kids. Greedy people know no such satisfaction. They work whenever the benefit exceeds the cost and amass as much as possible. They see no finish line and set no absolute goals. They seek only relative victory, and their desires are simple: get more than everyone else. Don't keep up with the Joneses, bury them.

When the inevitable bad times come in the form of starvation, drought, or disease, who is more likely to survive? Who were our ancestors? Who are we? As long as extra resources improve the chances of survival, people with materialistic drives and behavior will dominate. We run on the happiness treadmill because we are their great-great-greedy-grandchildren.

Of course this begs the question: Do rich folks really have more kids? Well, Bill and Melinda Gates, the world's richest couple, have only two, and they seem representative of a world where poor people have more children. There's a problem, however, in trying to understand the relationship between resources and reproduction by looking at the *current* relationship between these traits among people in modern industrialized countries.

The genetic evolution of greed, just like that of other behaviors, depends on ancestral circumstances, not on those of Bombay, Berlin, or New York City in 2000. In other words, the more meaningful question is: Did "rich" humans have more children than "poor" ones in our evolutionary past? Historically, did babies and riches go hand in hand?

Without a time machine we can't know for sure, but we can observe people throughout the world living in conditions similar to those of our ancestors. In many cultures, wealth does indeed translate to babies. In one study in the African country of Gambia, for example, women who were given extra resources had more babies. These women, like many people, were hungry. Extra food allowed them to feed their children and still have enough energy to become pregnant.

We find more support for the biological roots of greed by looking at the modern peoples most like our ancestors — living by hunting game and gathering plants. In these foraging societies, people are chronically hungry and their wants are real. Those who attain more wealth, especially in the form of food, have more children. So their desire for more has dramatic and obvious evolutionary consequences.

For our ancestors in a harsh world, greed paid off in the only currencies that matter to genes — survival and the ability to have offspring. From them, we have inherited a greediness that manifests itself today as a desire to accumulate money and possessions. So even though wealth may not relate to babies in an industrialized world, our instincts come from a time when concerns over material possessions were crucial.

It is, of course, possible to rein in these desires. The Indian leader Mahatma Gandhi, for example, survived with only a handful of goods, including his clothes, a pocket watch, a pair of glasses, and a walking stick. Few of us are as strong, so we run on the happiness treadmill with literally thousands of possessions.

⤵

Progress. Imagine going to the post office to mail a package during the holiday season. As you enter the crowded scene, you must choose one of two lines. The first choice is an hour's wait in a short line that moves slowly. Each customer takes forever, as the clerk keeps leaving to find a supervisor for help with complex details. The second choice is an hour's wait in a long line that moves quickly. People ahead of you zip past the clerk, and, as you wait, the pace picks up so that you practically jog up to the counter.

Which line would you prefer? For most people, the second line is much better even though the waiting time is identical. Most of us are pleased by two features of the longer line: it moves faster, and the rate increases over time. Several studies show the premium we put on such progress.

In one study, volunteers were paid to place their hands in icy water. One hand was held in bitterly cold water for sixty seconds. After a break, the second hand was put in exactly the same condition, plus an additional thirty seconds of pain was tacked on. During the additional thirty seconds, the temperature gradually rose from really cold to just darned cold.

The subjects were then asked to choose one of the two experiences for a third session. Which do you think they preferred, sixty seconds of pain or ninety seconds of pain? Overwhelmingly, they chose more pain. This seems puzzling because the longer version is exactly the shorter version plus added pain. However, the longer version ends with a positive trend.

A related study observed men undergoing a colonoscopy. In this unpleasant medical procedure, a relatively thick, inflexible metal tube is inserted into the rectum to view the intestines. For half of the patients, the exam ended the standard way — as soon as possible, the doctor ended the pain by removing the scope. In the other patients, the tube was left in place for some time after the exam. This stationary insertion is painful, but less so than the active examination.

Afterward patients rated their experience. Those with the longer procedure rated it as less painful overall, so the doctors predicted these patients would come back at higher rates for remaining treatments. As with the ice water study, colonoscopy patients preferred additional minutes of discomfort in order to have a better ending. We prefer experiences that conclude on a positive note.

In addition to progress, expectations are crucial to our happiness. How often have you gone to a movie filled with anticipation, only to be disappointed? Conversely, when you give a gift, do you ever try to increase the impact by downplaying it, saying that it really is nothing much? The saying "Satisfaction equals performance minus expectations" captures the central role expectations play in our emotions. Happiness and sadness derive from the difference between what we predict and what we get. This is true of experiences from movies all the way to life-threatening situations.

One Day in the Life of Ivan Denisovich follows a Soviet prisoner consigned to a Siberian labor camp. Although the book is officially fiction, the Nobel laureate Alexander Solzhenitsyn wrote from personal experience. Our hero, Ivan, has a truly awful day. He is starving and only gets a small portion of thin soup with bread. His clothes are threadbare, yet he must perform hours of manual labor in freezing weather.

As he goes to sleep, Ivan is "fully content" and concludes, it has been "a day without a cloud. Almost a happy day." Why is he happy instead of depressed? He's been in the camp for some time, so he fully expected the bad food, the hard work, and the cold. Because these horrible features have been fully anticipated, they caused him little pain.

Ivan's joy comes from several minor but unexpected breaks. He successfully steals a bit of extra food, smuggles part of a hacksaw blade back to camp, and acquires some scarce tobacco. For us, this day would be miserable. For Ivan, however, it soared high enough above his low expectations to make it nearly perfect.

Positive surprises make us happy, even when they are small. In one study, researchers gave some people a tiny gift and measured its effect. Specifically, half of the people using a photo-copier found a dime that had been planted in the coin return. After copying, the individuals rated how happy they were with their *entire lives* on a 7-point scale. So how much do you think a dime would increase a person's lifetime satisfaction?

Ten cents, if unexpected, buys us an enormous feeling of well-being. Those people who found a dime rated themselves a nearly perfect 6.5 — almost a full point higher than the 5.6 average of those who had not found any money.

Happiness and unhappiness are tools created by our genes to further their goals. Regardless of our circumstances, our instincts squeeze the most out of us. We are therefore very attentive to small changes that indicate progress and almost completely unmoved by anything that we expect. This efficient system makes us robust workers. With it controlling our moods, we can be set back, but we cannot be stopped. We dust ourselves off after defeat and look for ways to move forward. Our genes reward us with happiness whenever we make progress.

↜

Engineering happiness. In a folk tale, a troubled farmer seeks advice from a philosopher. He laments, "My house is too small, we are too poor to afford a larger one, and my family is at each other's throat from the close quarters." "Yes, I understand," says the philosopher. "I want you to go home and move your goats from their corral and into your house. Come see me in a week."

One week later the farmer returns looking even more haggard. He says, "My house is filthy and more crowded than ever. My daughter nearly killed my son, but fortunately was stopped when she tripped over a kid." "Yes, I understand," says the philosopher. "I want you to go home and move your cows into the house." This cycle of complaint and odd response continues for weeks until all the family's livestock is in the house.

Finally the philosopher says, "Put all the animals back outside and see me in a week." The farmer returns and says, "Yes, I understand. Our house is huge. We don't know what to do with all the space."

This fable highlights two truths about human happiness. First, change for the better produces joy, regardless of absolute levels. The farmer's house is exactly as crowded as it was before, but it feels much larger. Second, we can consciously structure our lives to make ourselves happier without any change in material circumstances.

We can start by recognizing the quirky ways our brain creates happiness and then capitalize on that knowledge. There are three important features of our genetic system. First, absolute levels have little effect on happiness. Second, we love making progress. Third, expectations play a central role. To be happy, we should therefore structure our lives to be on the upslope as much as possible. We should create situations and expectations such that surprises will be positive.

Giving gifts to friends and lovers is a fantastic way to capitalize on our instinctual happiness systems. Giving ourselves an extra meal or fine bauble is fun, but it's tough to surprise ourselves. In contrast, we can easily surprise our friends and make them much happier.

To create maximum happiness, we should give smaller gifts more frequently and tie them less directly to holidays and birthdays, when they are expected. Suppose we are planning to spend $100 on a birthday gift for our spouse or friend. We can create more joy if we divide that $100 into a $80 gift on the birthday and two unexpected gifts costing $10 each. As with finances, the key is to create as many positive surprises as possible.

These lessons also apply to physical fitness. Many of us begin exercise programs, then quit after a few months. We are motivated by rapid gains in the early weeks, then get tired as our ability plateaus.

To maximize our pleasure from physical activity, we should spend as much time as possible in the zone of rapid progress. This can be aided if we create our own "seasons," when we pursue one sport for several months, then switch to a new one. Just as we plateau in one activity, we move to another and again enjoy the early increase in ability. Similarly, taking some time off from our routines will help us stay "hungry" for more. Combining these two features, we can almost always be working on the steep part of the curve, enjoying the benefits of maximum improvement.

To take advantage of our thirst for progress, we need to divide big jobs into digestible chunks. During his doctoral program, Terry worked on a series of daily goals he put down on a list. While he was admirably successful in many areas, his thesis languished.

Success came after Terry realized that along with trivial tasks, such as "buy pants," he had listed "write Ph.D. thesis." While he was feeling good by accomplishing all his tasks except one (write thesis), his progress toward his most important goal was slow. Working on his thesis did not feel like progress because the "write thesis" task could not be crossed off his list after just a few hours. With a closet full of new pants, Terry learned to divide the big tasks into manageable subtasks so he could exploit his instinctual love for progress.

Ernest Hemingway was a keen student of human nature, and he manipulated his goals to increase his performance. Like many writers, Hemingway found that the toughest part of writing was getting started each day. He developed a practice of stopping each day with a chapter almost, but not quite, complete. The next morning he would want to get the little emotional reward we all get from progress, so he eagerly sat down to finish the chapter. Once immersed in work and feeling good, he usually continued writing.

The maxim of "under-promise and over-deliver" highlights the critical role played by expectations. Whenever we initiate a relationship or project, we should set appropriate expectations. The under-promise recipe is also useful when we fall behind schedule. An almost trivial example will illustrate.

If we are running ten minutes late to meet a friend and call to warn her, we should overestimate our delay, saying that we will be twenty minutes late. When we arrive only eleven minutes late, we are now early. Our friend will have a pleasant surprise to go along with a negative. When we give the initial bad news we take a hit, but we gain overall because people value positive surprises so much. We should always manage expectations so that we can exceed them.

Happiness requires continual effort. When we dream of our perfect world, we fill those images with laziness and indulgence. We imagine an endless stream of margaritas on the beach, limitless shopping trips, and days of watching football broken only by the occasional pizza delivery or beer run. Perhaps surprisingly, researchers have discovered that these sorts of activities do not make people the happiest.

Rather than ask people to *imagine* what would make them happy, an ingenious project determined what *does* makes them happy. Many times a day, the people in the study were beeped and asked to record exactly what they were doing and their level of happiness, answering a host of other questions in the process.

Success, not indolence, makes people happy. Specifically, the term *flow* was coined to summarize enjoyable situations. People experience flow when they are in control of their environment and using their skills to achieve a challenging and clear goal. In striving for these goals, we lose ourselves in the moment, become less self-conscious, and even have a sense that time has slowed down.

When do we achieve flow? Positive moments come in various domains, including sex and sports (as long as we feel proficient). One of the paradoxes is that people are more likely to experience flow while working than during leisure time. Oddly, even when experiencing flow at work, people imagine that they would be happier if they were not working.

So we think we'd prefer sipping gigantic beverages, yet we're happier progressing skillfully toward achievable goals. Our emotional systems are designed to encourage us to work. To further their ends, our genes have built us with relentless appetites for improvement and achievement.

In a memorable *Twilight Zone* episode, the main character starts in a hospital bed, then awakes to find himself in a hotel room. Whenever he wishes for anything, a bellhop appears instantly to provide the object of his desire. After some days of this, the man tires of his effortless existence and says to the bellhop, "I sort of wish I'd gone to the other place." "What place?" asks the bellhop. "Well, I assume I've died and gone to heaven, but I'm so bored, perhaps hell would be better." The bellhop responds, "This is hell."

ROMANCE AND **REPRODUCTION**

Gender

All animals play the mating game. In the movie *EDtv*, Woody Harrelson is caught on camera cheating on his girlfriend. When asked why he didn't turn down the woman's offer of sex, he says, "I'm a guy, I don't stop. That's the woman's job. . . . We're the gas, they're the brakes." What is the origin of our stereotypes for men and women?

Believe it or not, we can learn a lot about human sex stereotypes by examining an obscure little insect called the bush cricket. Although bush crickets don't have bars or dating services, they do congregate, check each other out, and — just like humans — ponder whether they should perhaps mate with a new acquaintance.

Ah, but there is one small difference. When the pair decides to mate and the male cricket ejaculates, he loses about a quarter of his body weight — contributing a massive ejaculate that the female uses for energy. For an average human male, this would be about fifty pounds of semen! If this were the rule in men, rather than the "lovin' spoonful" it actually is, would human males behave differently? The answer is yes, and the ramifications extend far beyond sexual tactics.

For a female cricket this sexy meal is important. The more nutrition she gets, in the form of food and sperm, the more baby crickets she produces. While we humans eat hundreds of times our weight in food over a lifetime, insects get by with much less. Their total lifetime food intake can be as little as two times their body weight.

A single ejaculation of a male cricket might, therefore, provide more than a tenth of the food that a female eats in her lifetime. Needless to say, this is a precious commodity, which she is careful not to waste. She efficiently converts nearly all of it into fertilized eggs.

As we might predict, male crickets are choosy when it comes to selecting a mate (wouldn't you be if it involved a fifty-pound ejaculation?). In particular, males reject small females that would produce few babies. With a maximum of a few bouts of love in his short life, a male cricket is intent on finding the mate that will best further his reproductive goals. Females, on the other hand, are looking for sex. After all, they can look forward to a nutritious sperm meal. Just as we'd suspect, females court males. If they could talk, we can imagine that the females would make all sorts of promises of fidelity and love.

In contrast to discerning male crickets, data do confirm that men are easy. Human males show little hesitancy to engage in casual sex. In a study of college students, 75% of men expressed a willingness to have sex when propositioned by a female exper-

imenter of average attractiveness. (Many of the men who declined actually apologized.) How many women said yes to the
same question asked by a male experimenter? *Not one.*

⌒

The gender question. Should women drive tanks in the Marine
Corps? Does it matter if a child's Cub Scout master or softball
coach is gay? Why are most politicians and racecar drivers men?
Why are nearly all grade-school teachers women? Have men
and women been conditioned by society to behave differently
or is something else going on?

Our culture has unquestionably had an effect on relations between the sexes. American women did not secure the right to
vote until 1920. Until Ronald Reagan appointed Sandra Day
O'Connor to the U.S. Supreme Court in 1981, one might have
mistakenly claimed that it was "natural" for men — and only
men — to be justices.

We know, of course, that the composition of the courts and
many other male sanctuaries is completely a function of cultural influences. Until 1984, men believed that women were too
frail to run an Olympic marathon. Similarly, many eighteenth-
and nineteenth-century female norms — including a complete
absence from the legal and medical professions as well as from
any leadership roles in religious organizations — ought to be
attributed to patriarchal restraints rather than innate differences between women and men.

Because women have been barred from so many activities, it
can be difficult to elucidate the genetic influences on gender
roles. Beyond basic plumbing, are there any aspects of male and

female bodies and behavior that we can be certain are caused primarily by genes? The answer is a definite yes.

～

We sure do look different. For one thing, men weigh 20% more on average than women, with the bulk of their extra weight in muscle mass. Consequently, men outperform women in most measures of strength. They are also an average of five inches taller. This does not prove that genes instruct male bodies to grow bigger, though; a lot can happen over the course of a life — independent of genes — to influence their height and weight.

If we look at younger children, however, we can filter out progressively more of the influences of our environment and upbringing. Ask some twelve-year-old boys and girls to throw balls as far as they can, for instance, and there is almost no overlap; the top girls throw only as far as the least skilled boys.

Of course, by the age of twelve, partly due to the urgings of their parents, most boys have played organized sports for years. So let's gather instead a group of two- and three-year-olds and see how far they can throw. Granted, these little tykes can't throw a ball very far, but even at this early age, before the influence of socialization is too overwhelming, 90% of the boys can throw farther than the average girl.

With ultrasound equipment it is now possible to extend these studies right into the womb. Fortunately for pregnant women, scientists haven't asked fetuses how far they can throw a ball. But they can make estimates. Male fetuses have larger arm bones — both the radius and the ulna — than females relative to their body sizes. In other words, long before boys have been

encouraged to start launching projectiles around the house —
before Mom even knows whether she's carrying a boy or a girl
— the boys have an advantage over their sisters.

Beyond these physical differences in weight, height, and muscle
mass, we can't be sure about the biology of gender. Nevertheless, there are some striking regularities across all modern and
historical cultures that are worth discussing. In almost every
society, women live longer than men; the average difference is
seven years. Is this difference just a cultural artifact — a non-
biological consequence of some feature unique to women's lives?
Probably not, considering how universal this finding is.

Among the native South American tribe of the Yanomamö, for
example, life is short and violent: even with an average longev-
ity of only twenty years, girls still outlive boys by almost a year.
Among Russians, the grim reaper works a little less efficiently
— life expectancy is sixty-five — but with a similar penchant
for males, who can expect thirteen years fewer than women.

Overall, 96% of the nations of the world report longer lifespans
for women. Interestingly, those few areas where men live longer
are mostly cultures that treat women terribly. In India, for instance, boys are fifty times more likely to be taken to hospitals
than girls, and girls are already four times as likely to be mal-
nourished as boys. So in India, at least, men live longer because
women are underfed and denied adequate medical care.

Just how big a role does biology play in the way men and
women age? A final bit of intriguing evidence comes from studies of the male hormone testosterone. In the early 1900s, men
committed to sanitariums frequently were castrated. Removing
a man's testicles reduces his testosterone levels to almost zero.

In perhaps the most gruesome example of the quantity versus quality of life trade-off, these castrated men — like neutered pets — lived much longer than similar, testosterone-laden men. How much longer? Well, it seems the cost of keeping your testicles runs about fifteen years! The same is true of all animals: one of the only sure-fire ways to increase a male cat's lifespan, for example, is to remove its source of testosterone. We can conclude that, just as men are built taller and heavier, women are built to live longer.

In searching for other human universals, we're confronted with the obvious and frightening fact that, throughout the world, men commit the vast majority of crimes. In the United States, for example, there are currently about two million people in prison; 93% are men.

With the invention and proliferation of guns — the great equalizers of crime — physical limitations no longer restrict women from lives of crime. Yet there hasn't been any dramatic increase in crime by women or in the use of guns by women. Among robbers, for instance, 50% of males use guns compared to only 30% of females.

There is also evidence that male and female brains function somewhat differently. After suffering strokes, women recover their language abilities more quickly than men. Now, brain-imaging techniques are starting to explain this ability, revealing more balanced brain activity in women's brains than in men's.

Behavioral tests have also long documented that men and women use and interpret language differently. Women can, for instance, name objects and articulate words more quickly than men. If asked whether two nonsense words rhyme (i.e., "gooz" and "rews"), women again outperform men.

In another study, men and women were asked to reflect on the saddest images in their lives. As they did, scientists used brain scans to monitor which parts lit up — a reflection of increased mental activity. In both sexes the limbic system — that part of the brain associated with expressing and feeling emotion — glowed brightly. In the women, though, the activity covered an area eight times larger than that in the men.

Genes build men and women with different bodies, and our brains have some subtle differences. Does this mean that genes build us to have different brains? Not necessarily. Take language. It could be that brain differences cause men and women to process language differently. But it could also be that girls are encouraged to express themselves verbally and that brain differences develop because of differential language use. In either case, though, the differences are real and intriguing.

In summary, physical differences and the strong cross-cultural regularities in certain behaviors suggest some biological foundation to our gender roles. Because of the long-standing and pervasive oppression of women, however, there are many unresolved questions. Freed from gender discrimination, for instance, would female crime rates approach those of males in a manner similar to the hoped-for convergence in salaries?

Let's return to our animal friends to see if their behavior, in a world free from TV and other cultural influences, can help us understand our own lives.

⌣

Gender roles in animals. Female Saharan gerbils have got it pretty easy. They settle down in the desert close to good food sources and don't travel much. Once they find a nice spot, they build a home with multiple exits so they can elude predators.

The life of male gerbils is less idyllic. They spend their time running across the dangerous desert sands, calling on as many females as possible. Arriving at a female's home, a male dispenses with pleasantries and inquires whether she's interested in sex. If her answer is yes, a brief bout ensues. If not, the male leaves unsatisfied. In either case, the male is back on the trail again within minutes.

It's tough to be a male gerbil. Whereas females can just set up house in prime, safe locations and concentrate on bulking up on food, males are on the road all the time; they eat less and are eaten more frequently. Consequently, a male gerbil's life expectancy is much lower than a female's. Equal numbers of both sexes are born, but you'll find at least twice as many females in the desert.

Why do males live such high-risk lives? Well, you can't make sales if you don't make calls. Imagine what would happen to a male who decided to forgo this perilous mating game. He might live to a ripe old age but would leave no offspring. Sons inherit genes from fathers who find the dangerous treks across hot desert sands worth the effort.

Female gerbils, it turns out, don't live completely carefree lives either. After giving birth, they get no help from males. In fact, that's another way of understanding the male's behavior. Because females do all the work of raising babies, a male who happens on a willing female gets a genetic reward with no further effort. It's this payoff that motivates such dangerous travels.

The same theme of male risk-taking to secure female child care occurs throughout the animal kingdom. In Natterjack toads, for example, males produce booming calls to attract females. Because females desire large males and size determines volume, females push their way through murky swamps until at last they reach the loudest croaker. They are drawn by calls that can be heard a mile away and are often louder than the legal limit for a car engine.

What do males get for their calls? Well, sex, of course. Quiet males are completely ignored by females. But toads that make noise are also at high risk to be eaten by bats that hone in on the calls. Male toads that want sex must take risks.

So in gerbils, toads, and thousands of other species, we find a common theme: females do the bulk of the work when it comes to offspring while the male contribution is just a bit of sperm. Accordingly, females hold the keys to the bedroom, and males have evolved to compete for their favor. With such a grand genetic prize at stake, the males take deadly risks to win.

~

Mr. Moms of the animal world. So far we've seen males compete for females, but that's not the only game in town. Among some tiny birds called phalaropes, for example, everything about the sex roles is reversed. Females are 25% bigger than males, and it's the females that maintain territories containing one or more males. The females court males and will even kill another female's chicks in cold blood if it helps them gain access to a male.

What causes this behavior? In this species, the males are the ones that care for the babies. A male can take care of a maximum of four eggs, but a female can be an egg factory, laying up to four times her own body weight in just forty days. Females find a guy, mate with him, lay four eggs, and then hit the road in search of more males.

The sex roles are reversed, but the message is the same: one sex provides most of the parental care and the other gets a free ride on the investing parent. Accordingly, members of the freeloading sex, that don't invest much, compete with one another for genetic prizes.

The "Mr. Moms" of the animal world uniformly demonstrate these behaviors. In one gender-bender case study, we meet the moorhen. One sex is big, aggressive, and nasty — they're the females. The other sex is soft-spoken and diffident. These are the males. The much larger females aggressively squabble over mates who will incubate their eggs.

Furthermore, the moorhen females are particularly picky about male body aesthetics. The ideal father must have large energy reserves on his body but still have the small frame optimal for incubating eggs; in other words, he's a squat, pudgy machine perfect for sitting on eggs. Females get all worked up over these boys and mix it up in vicious fights for them.

As we continue our tour of animal mating behavior, we return to the bizarre and frightening elephant seal. During the breeding season, these animals appear on islands off the coast of northern California. The females hang out in large groups on just a few prime beaches. Because they stick close together, the biggest males can dominate others in a competition for sexual access. In one study with 115 males, the five highest-ranking males fathered 85% of the offspring.

Among elephant seals, a male's life is short and filled with violence. With rich genetic rewards at stake, males are built for battle: more ferocious than a charging bull and three times the size of a female. During the three-month mating season, these powerful animals guard parts of the beach and do not take even a moment to eat. Even if a male manages to attain top rank for a season, he is often so exhausted from lack of food and constant battles that he is never seen again after his reign of glory. The vast majority of males die without ever having sex.

Body size is an important clue to behavior. In contrast to the enormous difference in body size of elephant seals, males and females of most bird species are the same size. In these species, even expert bird-watchers often cannot distinguish between the sexes except during pregnancy. Why do species differ in this way?

One rule of thumb for predicting the size differences is the level of competition. If one sex has a winner-take-all tournament, then members of that sex will bulk up to win the competition. One clue to the competitive nature of a particular species, therefore, is counting babies.

In elephant seals, a male may have over 100 offspring, compared to the female record of 8. In contrast, male and female kittiwake gulls have almost identical numbers of offspring. The documented lifetime record for a female gull is 28, for a male, 26.

Frequently, the level of competition in a species hinges on how helpless the babies are. In many bird species, it takes two parents to raise a brood, so a pair may stay together for a season or longer. If each female can only pair up with one male, there really can't be much competition for mates.

In turn, we can predict a great deal about the parental practices of a species just by looking at a picture of a male and a female. Simply examine the ratio of body sizes. If the two are dramatically different, like moorhens and elephant seals, you can be confident that the smaller sex is doing most of the childcare.

So far, our discussion has focused on the ardent sex. But the story is more nuanced. Male moorhens do not accept any female. They don't have to. Besides, they are interested in getting the best mate possible. Since males will bring up the family, they want the best genes in those eggs.

Learning about ourselves by watching animals. Can our knowledge about the sex lives of animals help us find meaning for our own lives? In many respects, we are simply average mammals. A woman carries the physical burden of pregnancy and, until recently, provided several years of milk for the baby.

The total investment of a woman in one pregnancy is estimated to be 80,000 calories, or the equivalent of more than 300 hamburgers at McDonald's. Better still, think of it as the amount of energy needed to run 800 miles. (Are you really willing to run from New York to Florida for a night of sex?) In contrast, a man's investment may not even last as long as a Super Bowl commercial or involve more than five milliliters of fluid.

The 1975 movie *A Boy and His Dog* illustrated an even more dramatic difference between men and women. It is the year

2024, and nuclear war has turned the earth into a barren desert. One survivor is played by a young Don Johnson, who struggles to make a living with the help of a canine pal. In a key scene, he is kidnapped by a group of survivors living underground.

Years of inbreeding in this society of about 100,000 people has "thinned their seed." The kidnappers reveal that they have captured our young star so that he can impregnate the women and rejuvenate their tired genetic stock. Envisioning a future even brighter than starring on *Miami Vice*, Don is instead soon horrified to find himself strapped into a sexual milking machine. At his expense we are reminded that a tablespoon of human semen contains enough sperm to fertilize every woman in North America.

They couldn't have made *A Girl and Her Dog* because women are limited to one pregnancy every year or so at most. The women's record holder is Mrs. Feodor Vassilev, with 69 children (in 27 pregnancies); Emperor Moulay Ismail's harem produced 888 children. It seems a fitting reminder of the pervasive suppression of women that no records even exist of the first name of the female reproductive champ. She is doomed to be remembered simply as Feodor's wife.

Our animal evidence leads us to predict that human males would be larger than females, that females would be less likely to die from a range of behaviors, and that males would take part in risky competitions designed to attract females.

How do these predictions fare? Well, as we noted earlier, human males are 20% larger than females and men do die at younger ages. For every mile driven on American highways, men are much more likely than women to die in an accident. The biggest divergence comes in the teenage years, when young men die at more than two and a half times the rate for women.

Other animals show a similar risky pattern. Among a species of primates called macaques, for example, male mortality is higher during the years that rank is being established. But it's a risk worth taking: high-ranking males have preferential access to females. Other animals show seasonal variations in risky behaviors. In one group of rhesus monkeys, for example, male injuries from fights soared by more than 600% in the mating season.

We can see other evidence of these differences in human behavior. There is a huge difference, for example, in the market price for male and female gametes. For a sperm "donor" the going rate is about $100 per ejaculation. Human eggs, on the other hand, fetch from $5,000 to more than $80,000.

In addition to a difference in willingness to have casual sex, there are other regularities in the commercial sex trade. With the exception of a few Chippendale men, exotic dancers are women. Pornography is also used primarily by men. (Even a survey conducted by *Playgirl* found that men were the ones looking at the nude male centerfolds.) Myths of the American gigolo notwithstanding, male and female prostitutes share one feature: their customers are male.

↪

Are there genes for homosexuality? How does evolutionary biology deal with a behavior that seems at odds with reproduction? Natural selection certainly seems to place a premium on having babies, and exclusive homosexuality appears an unlikely route to reproductive success. The short answer is that no one has a satisfactory explanation for the prevalence of homosexuality in humans and other species. Still, there are some interesting hints.

First, we know that homosexuality has a strong genetic component. Scientists often examine twins to try to untangle the role of genes. Identical twins have the exact same set of genes, whereas fraternal twins share only about half of their genes. Traits like eye color are determined by genes and are always shared by identical twins. Other traits like height are influenced, but not completely determined, by genes. So identical twins are closer to each other in height than are fraternal twins.

One study looked at sexual orientation in 55 pairs of identical male twins and 55 pairs of fraternal male twins. In each pair, one of the twins was known to be gay. The researchers sent a questionnaire to the brother, asking about his sexual orientation. Among the identical twins, in 52% of the cases both brothers were gay. Among the fraternal twins, in only 22% of the pairs were both gay.

Second, human homosexuality appears to be affected very little by childhood influences. One researcher looked at a sample of families where sons had "girlish" habits. A quarter of the parents were distraught enough that they enlisted the help of psychologists trained to discourage homosexual tendencies. Did it work? Not one bit. As adults, three-quarters identified themselves as gay or bisexual, actually a slightly higher percentage than those boys who received no counseling.

Further evidence of the relative unimportance of a child's environment comes from some societies in New Guinea. In an area called the "semen belt," tradition demands that all young males engage in homosexual acts, teaching that the consumption of semen is required for a boy's growth to manhood.

The best studied of these societies, the Sambia, are among the most warlike of all cultures. After their adolescent years of exclusive homosexuality, most adult men marry women and become exclusively heterosexual. In fact, despite the childhood indoctrination, the prevalence of adult homosexuality among the Sambia is lower than that in the United States.

Finally, if we look beyond humans, we find that same-sex stimulation takes place in many species. Our close genetic neighbors the chimpanzee-like bonobos are particularly enthusiastic participants in all sorts of sex.

Females pair up frequently for what is called g-g rubbing, in which they face each other and grind their clitorises together with 2.2 side-to-side moves per second, the same timing as a male during intercourse. Sometimes a panting and ecstatic female bonobo will even fall out of a tree and crash to the ground, stunned, when she becomes so caught up in her little pleasure session that she forgets to hold on to something.

In addition to same-sex contact, humans, apes, and some other large-brained species have separated sex from procreation. For example, bonobo adolescents — male and female alike — frequently perform oral sex on young males, and it is common for adult males to masturbate adolescent males. In humans, many heterosexual couples engage in sex that could not result in babies, such as sex during most of the woman's reproductive cycle, sex during pregnancy, and oral, anal, and masturbatory sex.

So the non-reproductive nature of human homosexual contact is shared with many other behaviors. What may be unique, however, is that many humans retain an exclusive homosexual orientation throughout their life. In contrast, many animals have homosexual contact only in specific situations.

Among gelada baboons, for example, the biggest adult males maintain harems of females. Smaller males that are left without any females, meanwhile, travel together in male groups with frequent homosexual stimulation. Once one of these males attains a female, however, he engages exclusively in heterosexual behavior.

Even these switch-hitting baboons lead boring lives compared with members of some species who can naturally change sex. For example, the blue-headed wrasse is a fish that lives on coral reefs. All wrasses begin life as females and continue to pump out eggs as they age and grow larger. Every reef section has one extremely large, territorial male that fertilizes up to forty batches of eggs a day. When the male dies, the largest female spontaneously changes sex and starts producing sperm.

Wrasses have simply pushed the concept of gender to its logical conclusion. As we have seen across the animal kingdom, one sex inevitably invests more in the production of offspring and the other sex competes for access to these choosy, committed parents. The greater the disparity of investment, the more dramatic the differences — physical and behavioral — between the sexes.

Beyond the question of *why* we see gender differences, there is the practical question of *how* these differences are actually brought about. Genes that underlie a behavior — say, mate choosiness or territoriality — can find themselves in a male or female body. Depending on which it is, the genes must often be expressed in dramatically different ways. Let's take a look at how they can do this.

⌐

Hormones induce many gender-specific behaviors. Young males are said to suffer from "testosterone poisoning." The label is appropriate: testosterone is a proven carcinogen, and, as we saw earlier, males without testicles live much longer than intact males. While all humans make testosterone, the level in men is about ten times that in women; it is one of the primary fuels generating male behavior.

We can learn a lot from people taking testosterone for bodybuilding and other athletic purposes because testosterone and a few of its chemical variants are so widely used. Look at some of the behaviors of men on steroids.

Gary had a friend videotape him while he drove his new Corvette into a tree at thirty-five miles per hour. On three separate occasions, Steve chased down the drivers of cars that cut him off in traffic. He then terrorized the drivers, smashing their windows with a tire iron. Chris rammed his head through a wooden door in a fit of anger. And Donny beat and almost killed his dog.

Their names have been changed to protect the stupid, but each of these moments of steroid-induced "glory" — besides being true — makes a case for testosterone as an important component of male madness. You see, none of these men had any history of violence.

Women, too, are powerfully affected by testosterone. On average, women with naturally higher levels of testosterone are hairier and have had more sexual partners. Among female prisoners, those with high testosterone levels are more violent. When women take testosterone supplements, they report increased confidence, added sexual desire and satisfaction, and are happier overall. Except for the minor fact that it can kill us and can cause violence, testosterone is a wonder drug that makes us powerful, confident, and happy.

Female hyenas have spectacularly high levels of testosterone. The result: they are larger than males and socially dominant. They also sport a pseudo-penis that — unless you're another hyena — looks just like the real deal. Nevertheless, female hyenas surpass the males when it comes time to care for the young.

Testosterone may be the hormone with the most dramatic effects on behavior, but it's not the only one. Estrogen, too, is a powerful force that can induce typical gender-specific behaviors. This was humorously — and perhaps a bit cruelly — demonstrated in experiments with rats.

First, some males were castrated at birth. Then, at puberty, they were injected with a little dose of estrogen. How did they respond? The confused male rodents immediately assumed the female mating position. In this stereotypical behavior — called lordosis — the male's front paws are lowered and his hind legs are raised, his back is arched, and his tail is moved to one side.

The obliteration of gender roles was made complete in this study when some females had their ovaries removed and were injected with a bit of testosterone. Sure enough, these females mounted other females and went through all of the thrusting motions of copulation.

The researchers, of course, couldn't resist bringing the experiments to their logical conclusion. Putting together a "male" that had been treated with estrogen and a "female" treated with testosterone, they created a bumbling — but apparently satisfied — pair of feverishly aroused animals in which an eager male in lordosis was mounted and "mated" by a female.

⌣

The road ahead. In a 1970s perfume commercial an Enjoli woman sang, "I can bring home the bacon, fry it up in a pan, and never ever let you forget you're a man, cuz I'm a woman." The advertisement reflected the gender politics of its time. Women, freed from historical shackles, could take on traditional male roles. If gender differences were entirely cultural, there should be no barrier to the complete homogenization of men and women. Some universities even went so far as to have coed bathrooms.

Today the picture is much less clear. Many traditional barriers to women's advancement have been torn down, but emulating males seems unlikely to emerge as the road to female happiness. Part of the reason is that so many of the things that make women and men happy are simply different.

With advanced technology, women no longer need to breast-feed infants. It's not much of a stretch to imagine a world in which pregnancy is accomplished by some technological feat, but even in that world, men and women would be different. Our genes come from an earlier era; our brains and emotional structures reflect that period and will not change anytime soon.

These biological differences complicate the push for equal rights. If men and women were identical, we might expect, or even require, equal representation in all jobs and in all majors at college. In light of our different tastes, however, it seems unwise to mandate equal outcomes. For example, the majority of young pediatricians and gynecologists are women. Does this mean male medical students are being discriminated against? Not necessarily.

Similarly, college enrollments are becoming increasingly dominated by women despite no apparent bias in admissions. Women get better grades in high school and appear to be more interested in getting college degrees.

So women are still from Venus and men from Mars. There is no simple path to ensuring gender equality. The hope is, however, that by combining equal rights with a deeper understanding of human nature, we can all be happier.

A polling firm asked, "If you were to be born again, would you rather return as a male or female?" Fifty years ago, most women wanted to come back as men and not a single man wanted to come back as a woman. In 1996, men still liked being men but women said they'd prefer to be reborn as a woman.

Beauty It's more than skin deep

Beauty is in the eye of the gene. What is beauty? At first glance the answer appears to depend on where and whom you ask. Among the Yanomamö of South America, for example, men use bright red makeup to advertise scars on their heads. Members of some cultures spend years enlarging their lips or stretching their necks. In many cultures, women never cover their breasts, while in others men spend inordinate amounts of time and money inducing women to uncover them.

Even within a culture the definition of "beauty" changes dramatically over time. In nineteenth-century America, feminine beauty involved pale skin and rounded bodies; current tastes lean in exactly the opposite direction. In the absence of any obvious common themes, we're left to conclude that beauty fads, like those in fashion and music, are controlled by the magazine editors, advertisers, and cosmetics companies of Madison Avenue. Right?

Wrong. The winds and whims of culture certainly play an important role, but beauty fads and fashions still rest squarely on a biological foundation. Consider the obvious, and practically invisible, connection between beauty and health. Who would you rather kiss, a person with clear, healthy skin or someone showing multiple symptoms of disease? A graceful athlete like Michael Jordan or a clumsy sluggard? Does a runny nose turn you on?

The answers seem like common sense, but where does that common sense come from? Was it necessary for your mother and father to teach you to be disgusted by open sores? Healthy, physically robust individuals are universally attractive, and this isn't an accident. Better genes are more likely to live in better bodies. We are descended from people who chose healthy, agile partners, and we have inherited their genetic standards of beauty.

We're no different from other animals in this respect. Female rabbits of some species chase males at high speed for long periods of time. Only after successfully running a mini-marathon is a male accepted into the role of father. Similarly, many snakes engage in vigorous male-female wrestling matches before sex. The meek may inherit the earth, but they will not be the descendants of slow rabbits or weak snakes.

Other sexual rituals have a health checkup built right in. Among the Ugandan kob (a relative of the antelope), all mating takes place while literally hopping across the African savanna. Males and females too clumsy or unhealthy to hop and hump at the same time are left behind in this mating game. Among many monkey species, intercourse requires males to balance themselves acrobatically on the hind legs of females. Not a feat to be attempted by the sick, feeble, or tired.

Humans across the world favor clear skin. As with the running rabbits, wrestling snakes, and hopping antelopes, there is a health reason. Parasitic infection, disease, and other illnesses often manifest themselves in the skin. A clear complexion advertises health among humans, just as vigor and strength do for animals. One study found that cultures where parasitic infections are common put an even higher value than others do on clear skin.

If we are interested in genetic quality, we're not limited to the skin. Good genes reveal themselves in other, subtler ways as well. As we mentioned in the introduction, one of these ways is physical symmetry — a feature that humans subconsciously find attractive.

If we were perfectly constructed, we would be exactly symmetrical. Why is that? Humans have a single genetic blueprint that specifies how to build both the left and right sides of the body in the form of "hand genes," "breast genes," and "eye placement genes." Individual genes do the work for both sides of our bodies. Departures from complete symmetry can be thought of as scars reflecting rough environmental conditions during development and a set of genes unable to cope with them.

No one, not even a supermodel, is built perfectly, though. We each have our deviations from left to right in arm length, foot width, breast size, everything. Generally they are quite minor, almost imperceptible. In one study, left and right index fingers, for instance, varied by 2%–4%. Marilyn Monroe was aware of her asymmetries. Of her two sides, she felt her right side was better. Consequently, it is very hard to find a picture of her taken straight on.

Across the natural world, there is a close relationship between an organism's symmetry and overall health and status. For example, among thoroughbred horses, symmetrical individuals run faster than their more lopsided competitors. Flowers that are more symmetrical produce more nectar, and bees preferentially home in on them.

Symmetry is also a near universal aphrodisiac. In a study of forty-one species, symmetrical animals were found more attractive and sexually desirable in over 75% of the cases. Life as a relatively asymmetric animal is a drag. Such creatures grow more slowly, die at younger ages, and have significantly less sex. With animals so keen to key in on this obvious barometer of health, should we be surprised if humans join them?

Of course not. Like other animals, humans care about symmetry, even if only subconsciously. One study presented people with a choice between a normal photo of a person and a photo of the same person manipulated so that the left and right sides were identical mirror images. The evaluators professed to be unable to distinguish any difference between the photos, yet when pushed to rate one more attractive, they overwhelmingly chose the perfectly symmetrical face.

In extreme cases, we can assess human symmetry with the naked eye. Most of the time, however, our guesses of who is symmetrical and who is not are wildly inaccurate. To really know requires precise scientific measuring devices.

Amazingly, we may actually be better at sniffing out symmetry. In one truly bizarre experiment, a group of men were measured for symmetry and given plain white T-shirts to wear for several days. They were not allowed to use scented soaps or colognes during this period. When the shirts were good and stinky, the men placed them in anonymously coded bags.

A group of women were then enlisted to open the bags, stick their noses in, and take a deep inhale of the aroma. When they recovered, they were asked to rate how much they liked each shirt. What happened? Across the board, the more attractive the smelly shirt's scent, the more symmetrical the man it had come from.

We're never really sure why — consciously, that is — but there is just something irresistibly desirable about symmetrical people. And it goes way beyond simply producing award-winning dirty laundry. Men with symmetrical bodies have sex three to four years earlier than other men and have more than twice as many lovers.

One study looked at elections and found, cross-culturally, that people can predict with uncanny accuracy politicians' electability by looking only at still photos and video clips of the candidates. Bill Clinton scores so high on symmetry that the computer rates him in the male model category.

Under the heading "More Information Than We Wanted to Know" comes a study that makes us wonder whether the men and women in lab coats with clipboards have gone too far. Over several months, the sex lives of eighty-six heterosexual couples were put under a magnifying glass. Both partners were measured for symmetry, then they were asked to record every intimate detail of their lovemaking.

The stunning result? The more symmetrical a man is, the more likely his partner will reach orgasm. Nothing else predicted the likelihood of females experiencing orgasm — not the man's attractiveness, not his height, not his potential earnings or his sexual experience, not even the couple's ratings of their feeling of love. (Right about now, we're all wondering where we can buy the equipment necessary for measuring our potential partners' symmetry.)

Why on earth would a female orgasm be influenced by male symmetry? The answer appears to be that this is yet another chance for a woman to exercise choice over who fertilizes her valuable eggs. When a woman has an orgasm during intercourse, more sperm is retained in her reproductive tract, and consequently she is more likely to become pregnant. Thus, differential orgasm rates improve a woman's chances for having a symmetrical child.

Cross-cultural studies offer more evidence for the biological basis of attraction. In many experiments, people are asked to compare photographs of people from other cultures and their own, rating them for attractiveness. For example, when American and Chinese men rate photographs of Chinese women, the rankings of beauty match.

Similar findings have been shown between numerous cultures, including India and England, South Africa and America, Russia and Brazil, and between black and white Americans. We agree on who is beautiful.

Do we all have the same conception of beauty because we watch the same TV shows and movies? To find out, anthropologists traveled far off the beaten track to visit the Ache and Hiwi, two isolated, modern foraging groups without any media contact.

When asked to rate photos for attractiveness, they find the same people attractive as every other culture does.

Finally, even three-month-old babies stare longer at pictures of attractive people than unattractive people. So underlying the vast variation in styles and beauty, we find biology and more biology. Beauty is as much in the gene of the beholder as the eye.

⌒

Who do we find attractive? Recently a Spanish couple about to get married made news when they discovered they were brother and sister. The general revulsion at their love reveals perhaps the most common feature underlying our relationships: our partners are not our siblings. There are few letdowns more stinging than to "be loved like a sister," meaning no romantic feelings. This feature of attractiveness is interesting because its biological roots are so clear.

Almost all animals avoid mating with close relatives because it makes for bad babies. From mice to monkeys, animals are reluctant to have offspring with siblings. If there are no other options, fruit flies will mate with a brother or sister, but they then show a noticeable delay in copulation as compared to mating with other flies.

The Russian royal family is a famous example of the costs of breeding with close relatives. It was filled with genetic disorders, including a prevalence of hemophiliacs. Babies who are born through incestuous unions die at twice the rate of others; those who survive show dramatically higher rates of many diseases, including mental retardation and heart deformities.

How do we avoid falling in love with our relatives? For most of us it seems obvious: we just aren't interested in our brothers and sisters. A study of marital discord, though, reveals exactly how this aversion develops. In some Asian cultures, marriages are arranged and future husbands and wives are brought up under the same roof from childhood. This is when things go wrong.

There appears to be a crucial period — from birth to two and a half years of age — during which couples that cohabit indelibly identify each other as sibling rather than spousal material. If they marry later, things rarely work out. The Asian couples brought together as infants divorce at three times the rate of other couples, for instance, and have many other problems, including an increased rate of infidelity by women.

Human incest avoidance thus has a clear purpose and mechanism. We avoid our siblings for the genetic health of our offspring. We have an instinctual, unconscious romantic aversion to anyone we grow up with at a very young age. Returning to our Spanish couple, it should be no surprise that this brother and sister had never met until adulthood; otherwise they'd have never fallen in love.

The desire for genetic diversity in our partners extends beyond simply avoiding our siblings. At a party a few years ago, Jean, a friend of ours with a Western European heritage, was powerfully drawn to Ali, a man of Middle Eastern descent.

That's putting it mildly. Jean describes her attraction to him as an animalistic and visceral need, completely at odds with her conscious evaluation that she loved her current boyfriend and

Ali lived in a different city. Although her rational side thought he wasn't a good match, Jean's animal side won: she immediately began an affair with Ali and broke off her long-term relationship that had been heading for marriage.

Jean's experience accords with studies showing that we are most attracted to people who have different immune system markers than our own. This biological feature, called MHC or HLA, varies among people and is more likely to be different in people from different parts of the world. When we mate with people very different from ourselves, we bring together combinations of genes that make for vigorous, healthy offspring.

Our discussion of the mating game up to this point highlights the tremendous amount that men and women have in common when finding partners. Both are looking for good genes in vigorous, clear-skinned, symmetrical partners who are not their relatives. Furthermore, as humans we are unique among the great apes in having both men and women contribute to childcare. We need each other and, accordingly, both men and women seek partners who are likely to be good parents.

This common ground is reassuring. But, as you're no doubt aware, the biological differences between men and women have some pretty big ramifications for behavior. A woman will produce about four hundred potentially fertile eggs in her lifetime. In contrast, a man launches three hundred million sperm per ejaculation. Because humans evolved in a world with few fertile eggs and a galaxy of sperm, men and women have a handful of different mating behaviors and aesthetic criteria.

What do men want? Look at the current Miss America and you will find one example of feminine beauty. Look back to the winner from twenty years ago and you will find another, quite different — perhaps thinner — example. Go back to Miss America from the roaring twenties for yet another, this one a tad plumper.

Indeed, a close inspection of Miss Americas reveals them to be a surprisingly diverse lot — in every category, that is, but one. Although the bodies of the winners are sometimes larger and sometimes smaller over the decades, their hourglass shape never varies. In particular, when the waist measurement is divided by the hip measurement for more than sixty Miss Americas from the 1920s to the '80s, the calculation never deviates from the tight range of 0.69–0.72.

What does this 0.7 ratio mean? What sort of woman can we conjure up with such a ratio? Well, one year it might be a 26-inch waist and 37-inch hips. A few years later it might be a 22-inch waist and 31-inch hips. Our perception of a woman's beauty, it turns out, depends more on shape than on size.

This desire guides more than just the Miss America judges. Consider this: Audrey Hepburn checked in with a waist-to-hip of 0.7 (31.5-22-31). So did Marilyn Monroe, at 36-24-34. This ratio holds for fashion models as well, from tiny Twiggy to Elle Macpherson, both coming in at 0.7. In fact, men from almost all cultures find women with the 0.7 ratio most attractive. When asked to rate photos or even drawings of women, this preference guides their choices.

Although subconscious, there is something special about that 0.7 that only a gene could love. Scientists studying conception found that women with the 0.7 ratios were the most fertile. A different study of women using artificial insemination found that women with ratios under 0.8 were twice as likely to get pregnant as women with ratios over 0.8. So men are attracted to a particular hourglass shape because it indicates fertility.

In addition, men throughout the world show a preference for younger women and for traits associated with youth — full lips, big eyes, and radiant hair. There's nothing specifically desirable about youth per se: it's simply that fertility tends to decrease as we age. Other cultural icons track these desires faithfully. With more than ten million units sold, "Totally Hair" Barbie, with a thick and flowing youthful coiffure, is the sales leader in this long line of dolls.

In a related line of research, computers are used to manipulate the images of female faces. The modified pictures are then rated for attractiveness. Oddly, an average face, a composite of a large set of photographs, is rated more attractive than almost any individual face. In part, this is because the merged photograph is more symmetrical than the individual pictures, but there also seems to be some beauty in the average.

The most beautiful women's pictures have some common features. They are more feminine, in the sense that they have thinner jaws, larger eyes, and a shorter distance between the mouth and chin. Cover girls share all of these features, as well as having plumper lips and smaller noses, all of which enhance the youthfulness of a face. So fashion models simply have exaggerated features of average women.

Many animals show this same behavioral preference. Consider herring gulls, for instance. Normally, when the baby birds peck at the red spot on their parents' beaks, they are given food.

When researchers made artificial parents by crafting wooden sticks with an even larger red spot than that on most birds, the chicks preferred the sticks over their parents.

Another researcher made fake fish that he called "sex bombs." They had grossly exaggerated fertility features compared to those of normal females, far beyond what would ever be natural. Male fish vigorously pursued the fake sex bombs and ignored healthy, real females. Our genes know what to look for, but if they're tuning into simple features they can be fooled.

Makeup and cosmetic surgery are attempts to enhance desirable features. Lipstick and collagen injections both make lips look larger and more youthful. Skin products seek to mimic the clear-skinned (remember, our genes read this as "parasite-free") look of healthy people. Because these preferences are innate, humans have used cosmetics since at least the time of the Egyptian pharaohs, and perhaps for as long as forty thousand years.

Jay Leno once commented, "In the last couple of weeks I have seen advertisements for the Wonder Bra. Is that really a problem in this country? Men not paying enough attention to women's breasts?" His point is a good one. Americans, among others, are obsessed with the female breast. Interesting theories abound, including the emphasis in shows such as *Baywatch*. To be sure, the media play a role, but this can't explain why, compared to those of other primates, human breasts are enormous.

Presumably there is a reason for this uniquely human feature, but what it is remains a mystery. There is no known reproductive advantage to large breasts. Women with larger breasts do not produce more milk or have healthier children. Nevertheless, we remain obsessed, and more than a hundred thousand American women undergo breast augmentation each year.

The male concern for physical beauty is not limited to women. Many studies document the high premium placed on looks and youth in the gay community. Compared to heterosexuals, homosexual men in personal ads disproportionately indicate a desire for physical beauty and advertise their own. Ads by and for lesbian women, in contrast, more frequently mention friendship and money.

One final attribute of attraction reminds us how simple desire can be: men like women who are interested in them. But how does a man recognize this interest? He may be tipped off by a seemingly trivial sign. When a woman is genuinely excited by a man, her pupils tend to dilate, sometimes becoming enormous.

In several studies, psychologists toyed with men's emotions by manipulating this usually honest signal. Men were approached and asked to participate in a psychology study (not realizing that they already were participating in one). Half of the female recruiters, it turns out, had their pupils enlarged with eye drops before approaching any men. The other half were working with their normal-sized pupils.

The result? Men volunteered in droves when the recruiters' pupils had been enlarged. Similarly, when photographs are altered to enlarge the pupils, men significantly increase their rating of the subject's beauty.

Significant benefits accrue to beautiful women in non-sexual domains. In a study of women in business, every point of beauty on the researcher's rating scale was worth $2,000 a year in salary. But there are costs to beauty as well. According to several studies, more attractive women are liked less by other women (even other beautiful women) and have a tougher time maintaining friendships.

What do women want? Marilyn Monroe reportedly dated John F. Kennedy for a while, but didn't have sex with him until he was elected president. In many species, females only mate with the highest-status males. Black grouse, for example, are plump, ground-dwelling birds that take part in a zany mating ritual. The males (cocks) congregate and vigorously fight for control of a patch of land while females (hens) sit on the sidelines and check out all the posturing and posing.

The territory that the males fight over has no resources, and neither males nor females feed there. It is simply a place for the males to show their wares. Because hens get nothing from the males other than sperm, it would seem that they are free to choose any cock that tickles their fancy. Yet they uniformly mate with the most vigorous male on his worthless territory.

Status pays for humans as well. In one recent study of almost two thousand marriages, women who married better-educated men had more success than others. Specifically, those with high-status husbands had more children, were less likely to get divorced, and were happier in their marriages.

All animals seek mates that help them in their quest for successful reproduction. Features that help in this struggle will be attractive, and women advance their interests by joining forces with high-status men. Have you heard about the powerful female senator who had sex with her young male intern? Neither have we. Great power, when held by women, is not an aphrodisiac.

In addition to status, money plays a central role in male attractiveness. In personal ads, women mention money more than ten times as often as men do. Women also advertise for love and commitment, characteristics conspicuously absent from the majority of ads placed by men. In psychology experiments, women strongly prefer ugly men wearing Rolexes to handsome men wearing Burger King uniforms.

When Jay goes to work some days, he wears what we call "the double Burger King." In fact, both of us divide our outfits into three gradations: "normal," "Burger King," and the aforementioned "double-BK." Although baggy sweatpants and a frayed T-shirt are the most comfortable, we remind ourselves that guys in low-status clothes are ugly.

Why, then, does Jay sometimes wear a torn T-shirt, shorts, and sandals? When he wants to work extra hard, he knows that he will be so embarrassed by the double-BK look that he won't leave his office. Consequently, he is not tempted to socialize or even go to any campus restaurants.

Women value resources in relationships for the same reason they value status — they further their own agendas with the food, clothing, shelter, and other goods that money buys. Interestingly, women with the best earning prospects place even higher importance on a potential spouse's financial position than do other women.

These female preferences show up in marriage data. Since it takes time to acquire resources, male attractiveness increases with age. In the average twentieth-century American marriage, husbands are three years older than their wives, exactly the same as in seventeenth-century Holland. Hundreds of cultures have a similar age gap.

Women are looking for financially secure guys with status, but they are not looking for old men. In the United States, however, forty-year-old males make, on average, $21,000 a year more than twenty-year-old males. So women seeking good mates end up with older men, but age itself is no turn-on. This explains why American men spend over a billion dollars a year to mitigate baldness, one of the most obvious signs of aging.

Women have a good genetic reason to avoid old men. Whereas a woman makes her full complement of eggs even before she is born, a man makes sperm throughout his life. As a man ages, his sperm increasingly become copies of copies of copies, with a consequent multiplication of DNA errors.

One night Kristi, a friend of Terry's, was unenthusiastically discussing her date to the college prom. Curious about the source of her displeasure, Terry asked a series of questions. Was her date smart? Yes, Einstein-like. Athletic? Muscles a-bulging. Witty? Razor sharp. "This guy seems like a great date. What's the problem?" asked Terry. "He's really short, like five feet tall," answered Kristi. Puzzle solved.

So in addition to status and money, a man looking for a woman should consider elevator shoes. Society grants all sorts of benefits to tall men, and not surprisingly, women place a high value on a man's height. In business, an inch of height is worth more than $1,000 a year.

Thirty-nine of forty-one American presidents have been above average in height, and the taller candidate has won almost every campaign. One notable exception was when the shorter Nixon defeated McGovern. Because there are so many benefits to being tall, we all posture a bit, and almost three-quarters of people exaggerate their height when asked.

~

How does our understanding of beauty help? Aesthetic values are derived, in part, from our animal heritage. They color our perceptions of mates, friends, co-workers, and politicians. Since we generally can't make ourselves taller or choose the genetic components of our sweat, does this knowledge help or hinder? We think it helps in two ways.

First, there's no guessing about what makes people attractive. As adolescents, we searched for magic clues to desirability. Did we need to become skinny, drug-addicted rock stars or emulate TV idols to learn the rules of the dating game? Now the answers are clear, and the good news is that many of these universal aphrodisiacs — which we'll elaborate on below — can be attained by everyone. Second, by becoming aware of our beauty biases, we have more control over our behavior.

Let's address these biases. Every human comes equipped with an unconscious preference for certain traits. In one study, for example, four hundred teachers were each asked to evaluate one fifth-grade student on a variety of scales. The teachers received the exact same written description and a photo of either an attractive or unattractive child. They rated the good-looking child more sociable, more popular, and even more intelligent.

Some people learn to use their looks to manipulate others. During a fifth-grade lunch, Terry was very surprised when Nancy, one of the cutest girls in his elementary school, came and sat with him. She said, "That brownie is stale. You're not going to eat it, are you?" Still stunned at his good fortune of having cute Nancy talking to him, Terry replied, "Of course not." "Good, then you won't mind if I eat it." Nancy took the brownie and never spoke to Terry again.

The phenomenon of granting favors to attractive women is known more generally as "the cute girl discount," and it's universal. In a slightly more scientific version of Terry's brownie experiment, dimes were planted in phone booths. When someone started using the phone, a woman would approach and ask, "Did I leave my dime there?" Nine out of ten attractive women got their dime back compared with only six out of ten plain women.

Lest we think that only bad people are biased in their behavior, recall that even babies stare longer at beautiful faces. We all have a preconscious beauty filter that influences our behavior toward people. If this favoritism seems as wrong to you as it does to us, we must try to adjust.

If we don't want to be biased by beauty, we have to set up preventive systems. One minor example will illustrate. Both of us require any students who seek a grade change to make their argument in writing. We know that if we allowed verbal presentations, we, like those fifth-grade teachers, would be influenced by our subconscious systems.

Returning to the basic question, isn't it unfair that we are judged by factors like symmetry that are beyond our control? Yes. But, happily, symmetry is only one of many features by which prospective mates will judge us. Fortunately, we have control over many of the other factors that will go into any mating calculation.

For example, one study asked men and women to rank thirteen attributes they would find attractive in a mate. The number one answer for both men and women was "kindness and understanding." Who among us can't improve on this characteristic every day?

Furthermore, while some commentators focus on the differences between men and women, we think the similarities are more striking. In the same ranking study, men and women both sought the same seven traits, including personality, adaptability, and creativity. Furthermore, except for men's placing a higher value on looks, men and women ranked these seven features in exactly the same order.

A final piece of good news comes from an unexpected source. As anyone who goes to a sperm bank will find out, women choose their sperm by reading about the donors. Usually there is no photograph of the man, but there is a lot of information. This is one area where we might expect people to be particularly ruthless in seeking qualities carved in genetic stone.

A group of Norwegian and Canadian women recently told researchers how they selected their sperm donors. They did indeed select donors for their physical attributes and good health. However, the features that influenced these women the most included honesty, dependability, and consideration. As with kindness, these are all features that we can improve on daily.

So we choose to see the beauty glass as almost full rather than partly empty. First, the road to attractiveness need not be a mystery. The behaviors of a good mate are almost exactly those that our mothers tried to teach us. (We would have listened more if we knew it would get us more than Cub Scout merit badges.) Second, anyone can become significantly more desirable by being considerate, staying physically fit, and being fiscally responsible. What could be fairer?

Infidelity Our cheating hearts

The specter of infidelity looms over us. Imagine for a moment your spouse or partner forming a deep emotional attachment with someone else — laughing together, confiding intimate thoughts, sharing long afternoons of companionship. Now imagine instead your partner having a brief sexual fling with someone else — impersonal but sweaty, loud, and visceral. Neither is a pleasant thought, but which scenario distresses you more?

Psychologists described these two scenarios to women and men and measured the resulting physiological stress. As expected, both are unpleasant for all people, but the thought of their partners having sex makes men more crazy, creating pounding hearts and sweaty brows. Women are calmer in both conditions but are relatively more threatened by romance. Let's investigate the deep biological roots of this gender difference, first by exploring marriage.

Finding a mate does not signal the end of tough battles. Far from it. Marriage and monogamy are not the Holy Grail for our genes. Among Americans, more than a quarter of women and nearly half of men admit to sex outside marriage. (Curiously, more than half of the women who read *Cosmopolitan* report infidelities.) Furthermore, while affairs begin earlier in the marriage than they did a century ago, the proportions haven't changed much. Who are these infidels and why is their resolve broken down?

A warning to husbands: the baby you hold may not be your own. When cheating leads to husbands who are not fathers, it is called cuckoldry. A British study found a cuckoldry rate of almost 10%. Estimates for other societies range from 1 to 30 percent. Using DNA fingerprinting, researchers examining 1,607 Swiss children found 11 cases of cuckoldry. Sadly, some men learn the truth when they volunteer to donate a kidney or other organ to their sick child. During tissue-typing before the surgery, it is sometimes determined that the man could not possibly have fathered the baby.

QUAYLE TO MURPHY BROWN: YOU TRAMP was the headline of the *New York Daily News* when the vice-president blamed TV for a decline in family values. From Jerry Springer to Ricki Lake, the airwaves are indeed filled with stories on the ethically impaired. Before attributing the infidelity figures to a breakdown in morality fueled by pop culture, however, let's check in with our animal friends outside the influences of movies and rock & roll.

In the absence of diamond rings and monogrammed towels, it's not always possible to tell whether animals are "married," but it is clear that most animals never form marriage-like bonds. Love 'em and leave 'em one-night stands are the norm. Monogamy among mammals, in particular, is rare.

Still, from Nile crocodiles to Australian dung beetles, plenty of faithful species have sprouted up around the world. Had Murphy Brown been a bird, she wouldn't have been a single mom. Male birds reliably play the good dad, bringing home worms and other treats for the brood.

Appearances can be deceiving, though. In crafty experiments, red-winged blackbirds have been caught in the act. Scientists with tiny scalpels and slightly crazed psyches can perform vasectomies on birds. The result is the same for birds as it is for humans: males can have sex and ejaculate, but they produce no sperm. Consequently, a vasectomized male (bird or human) has a right to be upset if his partner gets pregnant; in the absence of immaculate conception, a pregnant spouse has got some 'splaining to do.

The blackbirds in question spend the winter in the South and the summer in New England. Biologists captured birds migrating north, vasectomized some unlucky males, then released the birds and observed them throughout a breeding season. The vasectomized males fought and defended their territories as vigorously as the others. They also acquired partners for the season, mating enthusiastically.

Everything appeared to be in order: each female was observed eating, sleeping, and breeding with one male. Soon, eggs were laid and baby birds born. To the surprise of the observers, most of the females living with vasectomized males had babies. The existence of monogamy, it seems — and not just among humans — is less pure than we had imagined.

Faithful or not in life, human bodies are designed for infidelity. Begin with the observation that gorillas have golf ball–size testicles while chimpanzees' look more like baseballs. As a percentage of body weight, gorilla testicles measure 0.02 while chimps devote fifteen times as much energy to attain 0.30. Why the huge difference? Intrepid primatologists have investigated the amount of sperm needed for fertilization. The tiny testicles of the gorilla are perfectly adequate, but chimp males produce far more sperm than necessary.

Why do chimps waste the extra energy making big testicles and surplus sperm? Shouldn't evolution select for the least costly testicle that gets the job done? Efficiency is indeed the rule for gorillas, where all females mate only with one male, the dominant silverback. Fertile chimpanzee females, however, have sex dozens of times a day with many males.

Because chimps are promiscuous, the sperm of several males battle inside each female's womb for fertilization. The chimps who want to win these sperm wars must bring a lot of troops to the battle, hence the baseballs between their legs.

To learn the mating practices of a species, we start by examining the testicles. What's the answer for humans? We definitely aren't built for monogamy, but we aren't chimp-like tramps either. Human testicles average 0.08% of body weight, four times that of gorillas' but only about one-fourth the figure for chimps. Like chimps, humans produce more sperm than is required for fertilization — enough that men with only one testicle have little difficulty reproducing.

Furthermore, in humans about 99% of the sperm in an ejaculation are not fertile at all. Many of the non-fertile sperm are "seek-and-destroy" sperm that actively search for the sperm of other men and annihilate them, while others function as blockers, denying other men's sperm access to the uterus.

Would our bodies go to the trouble of building tens of millions of "anti-sperm" if there were no other sperm to battle? Moreover, after couples have been separated (and presumably both have had greater than usual opportunity to be unfaithful), men produce sperm with a higher percentage of the anti-sperm blockers.

Far from being a cultural creation, infidelity persists in spite of harsh social sanctions. Afghanistan, for example, has returned to Islamic law: a recent news report described a young man casting the first stone in the execution of his adulterous sister. In addition to stoning, cultures have tried "disincentives" that include whipping, permanent bodily mutilation, and public humiliation. Nevertheless, adultery occurs in all societies — rich or poor, Christian or Muslim, democracy or dictatorship.

Big balls and anti-sperm indicate that the prevalence and universality of human infidelity have deep biological roots. The question remains: Why? Newlyweds don't generally plan to cheat and often express regret after infidelities, yet in many marriages, and throughout the animal world, promises are made and broken.

↝

Marriage is an exchange. To understand infidelity, we have to understand marriage. Recall that women and men pursue their similar genetic goals through different means and different sorts of mates. In the romantic negotiation, men bring the promise of many things: time, commitment, caring, and cash. Women bring time, commitment, caring, and the promise of fertility. Both sides want what the other side's offering. Put starkly, marriage is an exchange.

Even in the toughest negotiations, two sides can usually find common ground. In the mating game, agreements lead to churches and wedding receptions. Many societies make the negotiated terms explicit, labeling marriage as the man's exclusive access to sex in return for his investment in his wife and children.

But genes never sleep. Instead of a blissful "they got married and lived happily ever after," gene fairy tales end with offspring and more offspring — any way the genes can get them. As surely as they drive couples down the wedding aisle in the first place, our genes will push us toward betrayal whenever infidelity is in their interest.

Infidelity is an attempt, often subconscious, by one side to improve his or her deal in the marriage exchange. Unfaithful women are seeking better genes for their babies and/or better partners. Unfaithful men are seeking additional fertility and/or better partners.

To investigate marriage further, we ask why people get divorced. The four F's — fertility, fidelity, funds, and sex — are the answer. Take infertility. Across all societies, couples with children are much less likely to divorce, and the more kids they have the less likely they'll split up. Many societies don't even recognize a marriage until children are produced.

Ring doves have a mating system based on fertility. Pairs form and remain together during a breeding season, but any relationship that doesn't produce baby doves is doomed. In a form of no-fault divorce, 100% of doves in barren couples match up with different partners for the next season. Pairs that successfully produce babies stay together year after year.

Even more important than fertility is fidelity. Although an empty nest can drive couples apart over the course of years, unfaithfulness can cause separation in just one night. In a study of 160 human societies, infidelity was overwhelmingly cited as the top reason for divorce. Other divorce patterns also reflect sexual issues. Contrary to the popular idea of a seven-year-itch, people are most likely to divorce in the fourth year of marriage. This four-year-itch is scratched across more than sixty radically different cultures.

Divorcing at four years allows wives and husbands to have their genetic cake and eat it, too. Both are likely to be in their prime reproductive years; they are still hot properties in the mating game. Not surprisingly, more than three-quarters of the people — women and men almost equally — remarry following divorce. Even as divorce rates have risen recently in the United States, our desire to be married has not waned.

Factors beyond fertility and fidelity also play into the divorce debate, especially money. After five years of marriage, a woman is three times as likely as a man to complain that her spouse is too stingy. Specifically, women complain about their husbands' lack of interest in buying gifts.

Jay mentioned this fact to an acquaintance, Russell. Practically yawning, Russell said "I don't need a couple of Harvard guys to tell me to buy more gifts for my wife." So Jay asked whether Russell actually gave his wife many gifts. "No. But I don't need you guys to tell me that."

Russell, start giving your wife more gifts. Today.

Divorce rates may also reflect female opportunity rather than a breakdown of family values. When women earn more than their husbands, divorce rates increase by fifty percent. Among the !Kung San hunter-gatherers, women gather the food, providing the majority of the calories. For these women, food translates to power and influence. It also translates into a higher divorce rate.

The same is true of North American Navajos and nearly all other societies with relative equality between the sexes. It may be that women all over the world have a relatively constant desire for divorce, but only the more independent can afford to act on it.

Males cheat, females fight back. The genetic influence on male infidelity is clear, well known, and can be summarized as "men are dogs." Male reproductive output increases with sex outside marriage, so the cost-benefit analysis is simple: males are tempted to cheat when they can get away with it. Most adulterous men aren't actually looking for a way out of their marriages: more than half of the cheaters even describe their marriages as "happy."

When are men most likely to stray? That depends on when they can find willing partners. Because male attractiveness is increased by wealth, men cheat most between the ages of forty and sixty, when they finally have the opportunity. These are their peak years of income. More than a third of the sex they have during these two decades comes with women other than their spouses. The men remain quite fertile and can continue to increase in status and power throughout these years.

Women are advised to "stand by your man." This is usually interpreted as a statement about loyalty, but it is also a good way to prevent him from cheating. Let's examine other behaviors that guard against this genetic disaster by returning to our original thought experiment.

What will disturb a woman most: her man's sweaty sex or his deep emotional bond with another woman? Is her reproductive success lowered if her husband spends a few hours, during a business trip, having sex with a woman he will never see again? Not necessarily. Deep emotional attachment to another woman, however, is more threatening, as it often signals the imminent end of the relationship.

In a world of wandering males, how can a female look after her interests? There is only one way a female can be absolutely certain of getting resources from a mate in exchange for access to her valuable eggs: payment prior to delivery.

If you are a female hangingfly, this system works just fine. Males and females alike prefer to dine on cuisines such as aphids and houseflies. They love to eat, but finding and catching these nuggets of nutrition can be tiresome and risky. The females solve this problem by linking food and sex. To put it simply, she won't mate unless her suitor brings her a big tasty treat.

When a male hangingfly — let's call him Hal — captures something worth eating, he may snack on it a bit, but then he holds it up and releases an airborne signal to all of the females in his vicinity. It's much like the irresistible smell of a turkey dinner on Thanksgiving. It proclaims: "I'm Hal, a great hunter. I've got good food ... for a price." Not much later, a female, maybe Miriam, shows up.

Hal presents the bug to Miriam (while retaining a tight hold on the bug) and then pulls her closer to him. (Life with six legs has its advantages; with the other two legs he may be simultaneously reading a book.) Hal then begins mating with his new friend and she agrees, continuing the mating for the twenty minutes fertilization requires. While Hal is getting his tiny rocks off, Miriam is eating that juicy bug as fast as she can.

When the twenty minutes have passed, Hal gets a little testy and takes the remainder of the bug back — he'll use it to attract another mate or eat it himself. Miriam, meanwhile, flies off sated and lays eggs. The exact number depends on how hearty a meal Hal gave her. If we're interested in romance, we'll have to look elsewhere. Nothing but a simple transaction here: food for fertilization. The more food given, the more babies produced.

Similar examples of "nuptial gifts" abound throughout the animal world. Hummingbird males, for instance, guard flowers from other males. A female needing nectar for fuel willingly accedes to a hangingfly-like deal: if the male lets her feed on nectar for a while, she will mate with him after she has filled her tank.

The award for collecting impressive payments up front, however, has to go to praying mantis females. They completely consume their mates (often eating the head during intercourse) and are perhaps unique in their realistic assessment of a male's promise to provide food for the family.

For our ancestors without birth control, sex meant babies, and a woman who had a baby without a caring partner was in serious trouble. That's why women demand a longer-term investment. Let's look at the courtship between our friends Kevin, a Wall Street trader, and Kate, an international model.

Although Kate politely declined Kevin's initial romantic overtures, his ardor persisted. Repeated dates ended with Kevin's dropping Kate off by limousine. He would say, "I'm going home, but I'll send the car back in case you want to come over." Wary of his intentions, however, Kate slept alone and would wave to the limo driver as she left for work each morning.

In some bird species, females demand that a male perform an elaborate and time-consuming courtship dance before she'll mate with him. A female demands that her beau court her long and hard before she turns over the goods. This courtship dance involves fancy dives into water, graceful hovering, flamboyant twists and turns, and generally his making a fool of himself for days on end. But after passing the grueling audition he can generally be counted on to stick around in order to see a brood through hatching and weaning.

The next best thing to an on-the-spot exchange is thus a believable pledge of commitment. But even as females demand promises, evolution demands that they be skeptical. Talk is, after all, cheap, and believing is risky. Accordingly, women seek proof when it comes to statements of love. Roses are nice, diamonds even better.

What happened to Kate and Kevin? After many months and numerous dinners, Kate was won over. After a point, Kevin had spent so much time and money conveying his interest that he had to be telling the truth. He was. They are now married with three kids, homes in four states, and a private island in the Caribbean.

⌐

Females cheat, too. Why do women cheat? Returning to the basics, access to more sperm is not the answer. After a few sexual bouts in a fertility cycle, additional sex will not increase the size of a woman's brood. More than one pregnancy per nine months just isn't possible.

Women have access to plenty of sperm; they needn't ever worry about shortages. Should it follow, then, that they be uninterested in opportunities for infidelity? Absolutely not. In the marriage contract, women are looking for good genes and commitment. By cheating, they may be able to upgrade one or both of these contractual terms. What this means, though, is that women will be considerably more discriminating than men in their infidelities.

We ought to expect women to have propensities for two different kinds of cheating: one that improves the genes in their offspring, and a second that helps them get more committed mates. They won't necessarily find both features in the same guy either. Zebra finches commit similar gender-biased infidelities; a gene-shopping female finch will only have sex with a male that is healthier or has a better territory than her partner, while a gene-depositing male will stray with any female.

Think back to the introduction. Remember our noting that there are four days a month when husbands ought to be especially attentive to their wives? Here's why: women who cheat on a spouse are most likely to do so during the four days surrounding their ovulation — their days of highest fertility! If you're shopping for genes, do it when you have room in your cart.

Though married women have only a small percentage of their sex outside marriage, their infidelities are concentrated in the fertile period. Furthermore, women are significantly less likely to use contraception with their lovers than with their husbands. This explains why many babies don't belong to "Dad."

But how do women know when they are fertile? Consciously, at least, they may not. But perhaps their hormones are influencing their behavior subconsciously. Nearly all female primates advertise their days of fertility with large and colorful genital swellings. A male chimp can spot an ovulating female from more than a quarter mile away.

Not so for humans. A man lying next to his wife in bed has no clue. Much to the dismay of couples trying to have babies (and teenagers trying not to have babies), we have been built with a rare system — concealed ovulation — in which the reproductive state of women is unclear. This system enables women to strategize against men, in part by allowing fertile wives to slip out and track down some better genes.

So some women cheat simply to get better genes for their children. Others use infidelity to secure better partners. Most adulterous women, in fact, describe their marriages as "unhappy," and three-quarters say they are seeking long-term commitment in their affairs.

Do children more closely resemble their mothers or their fathers? Psychologists have investigated this question by showing a group of photos to strangers, who try to match parent and child. For ten-year-olds, the strangers are equally likely to guess a child's father correctly as they are to identify the mother. For one-year-olds, however, people are much more likely to correctly pick the father. Babies resemble fathers because this induces males to care for them; confident dads change diapers.

⤳

Females use sex to help their babies. Women seek good genes and commitment from men. But that's not the whole story. Females sometimes use sex to modify male behavior toward that female's offspring. Why would they do this? Males in many species kill offspring fathered by another. There is a clear genetic benefit to this infanticide: by killing a young animal, a male may induce its mother to become sexually receptive sooner than she would otherwise. She is then in a position to have *his* offspring.

Scientists have studied male infanticide extensively among lemmings. In one experiment, males that had sex with a female were observed to almost never kill her pups. In fact, they often helped care for the young. In contrast, when a male was introduced that had never had sex with the mother, he killed, on average, 42% of her babies.

The exact mechanism for this infanticide was revealed by clever manipulation. Scientists isolated odor-causing chemicals from a young mother — call her Michelle — and rubbed them all over a female that had no babies — call her Nicole. Male lemmings were then allowed to mate with Nicole. When the males were later introduced to Mom Michelle and her offspring, they felt no compunction to kill the pups. The chemicals tricked them into believing it was Michelle they had mated with, so they acted just like loving dads toward her babies.

In lemmings we have a clear story. A male lemming uses chemical clues to determine which females he has had sex with, and he refrains from killing the babies of these females. A males acts aggressively, on the other hand, toward babies he knows are fathered by competitors.

The infanticide story is less clear in primates but shows some similarities. Among the langur monkeys of India, each troop usually has a dominant male that has primary sexual access to all of the females. Once every two years or so, however, this male is physically defeated by an outsider. The new male begins his reign by seeking out and killing the infants sired by his predecessor. He also has a tendency to kill any babies born soon after his reign began.

What does a female langur do when confronted with a new male? She has sex with him. Why? To further her own genetic interests, of course. If the male has killed her baby, it's time to start a new family, and he's the only guy in town. If she is already pregnant with the previous male's offspring, though, it's less obvious why she still seeks out the new male for sex. It's possible that the female uses this tactic to fool the new male into thinking her soon-to-be-born babies are his — and therefore leaving them alone.

Females in a variety of primate species appear to use sex similarly to manipulate males. Among Barbary macaques, for example, when females are maximally fertile, they have sex once every seventeen minutes for days on end. They aggressively pursue males and initiate sex at least once with every adult

male in the group. Again, no one knows for sure what the females gain by all this sex, but one theory is that the females are
altering male behavior. Males do not kill babies born to their
sex partners.

Women in some cultures use a related strategy to help their offspring. Among the native South American Ache, women openly
have sex with multiple partners, and a child may have more
than one man designated as the father. If so, one man is the
"primary" father and plays the role of a traditional dad, living
with the mother and providing for the baby.

What do women and men gain from this system? First, let's consider the payoff to women. Among the Ache, death often comes
early. If a woman's primary husband dies, one of her other husbands will step in and help both her and the child. In this harsh
and energetically demanding world, babies without fathers die
at much higher rates than those with a dad to help out.

Now take the male perspective. There are both costs and benefits to sharing a woman with other men. By sharing, a man decreases his chance of fathering a child. The benefit, however, is
that the children he does sire — because they have a backup
dad — enjoy a greater chance of survival.

During the most comprehensive study of the Ache, 63% of children had at least two fathers. This "insurance policy" holds significantly less appeal to males, of course, when their death rates
are lower — as in most industrialized societies. Indeed, those
males spend a great deal of time trying to ensure that they don't
end up raising another man's children.

〜

Males fear cuckoldry. When Nicole Brown Simpson was found murdered, O. J. would have been the prime suspect even without a bloody glove and Bruno Magli shoe prints. When a young woman is murdered, the savvy detective in any society starts by looking for an estranged male.

In an effort to retain female faithfulness, males use tactics ranging from passionate promises to surveillance, threats, and violence. Around the world and throughout history, homicide is one sad outcome of these male strategies. In 1998, one-third of the 3,419 women murdered in the United States were killed by men with whom they'd been romantically involved.

Return to our thought experiment at the beginning of this chapter. What disturbs men the most? The male fear is that the exclusive sexual access he has bargained for will be given to another. Unlike a woman, he can never be sure that the baby produced by his wife carries his genes. If cuckolded, he may pay to educate another man's child or perhaps even donate a kidney to this child. While emotional attachments between his partner and another man are threatening because they may lead to sex, nothing is worse than actual sexual infidelity.

Just as surely as genes play offense, they are prepared to mount defensive strategies. Because the blade of promiscuity is double-edged, a male must assume that other males will try to put the moves on his partner. He must stop this. But how?

The contrivances by which males keep their partners faithful — or at least try to — run the gamut from simple to extreme. At the simple end we find the reliable old standby of mate guarding. A high school girl is encouraged by her boyfriend to wear his varsity jacket or to display his school ring on a chain around her neck. Or he may brag to his friends about his relationship. These are simple signs of affection, to be sure, but read between the lines for the message from his genes to those of other males: "keep away."

Judith, a twenty-six-year-old graduate student, could be counted on to leave every party at 10:45 P.M. When pressed to explain, she admitted that her boyfriend on the other side of the country liked to call and chat with her every night at eleven. Like clockwork, fertile Judith would remove herself from rooms brimming with potential suitors. Was Mr. Boyfriend's goal simply love talk? Perhaps. But his scheming mate-guarding could scarcely have been more effective had he traveled to her apartment and sat on Judith every night.

Say, that's not a bad idea. In some species, males sit on females for prolonged periods after sex. Dispensing with jealousy, late-night phone calls, and varsity jackets, these males that don't want to raise some other guy's offspring do what they can to ensure a Father's Day gift that is genuinely theirs.

At its core, mate guarding, and the jealousy that incites it, are about insecurity and uncertainty. As long as the offspring emerge from the female's body, she can be certain that they contain her genes. In contrast, a male inhabits a "danger zone" that lasts as long as the female is fertile. If she mates with anyone else during this time, the child she produces may not be his. If he is going to help raise the offspring, he had better minimize his risk in the zone.

If a male wants to monopolize a female, why bother to stop mating at all? In many species, males take this approach to reducing risk in the danger zone. Among houseflies, even though the male has transferred his full load of sperm to the female in ten minutes of copulation, he doesn't let go of her for a full hour.

Moths put the flies to shame, continuing their mating for a full twenty-four hours. But the real champions are certain frog species that continue individual bouts of mating for several months. If people mated for a similar percentage of our lives, a single round of intercourse would last almost ten years.

Even if we've got the decade to spare, this strategy requires more than stamina and determination. Natural selection makes this all possible by building into the genitals of these males a ghastly collection of hooks, spines, and claspers that prevent decoupling before he is ready.

But it is only necessary to guard the female when she is fertile. Everyone from the elephant seal to the barn swallow seems to have keyed into this. Throughout a female's fertile period her mate keeps close, monopolizing her time. As soon as the season ends, though, this possessiveness wanes and he allows her to fraternize with whomever she pleases.

Anthropological studies on the island of Trinidad describe human mate guarding that mirrors that of other animal species. Men with fertile wives — that is, young wives who are neither pregnant nor nursing — spend a huge portion of their time guarding their mates compared to men whose wives are less fertile.

How do Trinidad husbands keep other males away from their spouses? No elaborate, elegant strategies here: they simply spend more time around the house. They also get into many more fights over her with other men. The easygoing husbands with less fertile wives get along famously with other men and spend much less time with their wives. Want to know how fertile a woman currently is? Get a tape measure and see how far her hubby is standing from her. (And get ready to back off when he asks what you're doing.)

In the absence of fertility data, if we want to predict how aggressively a man will guard his wife, we need look no further than age. Newlyweds or not, wives in their thirties are freer — by any measure — to do what they want than wives in their twenties. The inverse is not true. Women guard their mates equally vigorously whether the man is twenty, thirty, forty, or more. Women are wise to monitor their aging husbands; men in their forties are fertile, wealthy, and unfaithful.

Unless you've got a weak stomach, read on. In the world of anti-cuckoldry schemes, strategies from the drawing board of natural selection are more devious and grisly than those found in the creepiest horror movies.

Consider the weird ways of the black widow spider. Shunning the more traditional chastity belt, the male breaks off his sexual organ inside the female, preventing her from ever mating again. When the act is completed, the female kills and eats the male. *Femme fatale,* indeed. A rough justice prevails, however. In sealing his mate's reproductive tract, a male assures himself of fathering the spiderlings. In snacking on her lover's nutrient-filled body, a female gets the resources to produce those baby spiders.

There's no such fairness among the thorny-headed worms. With the development of "cement glands," males of this common parasite have elevated their protective tactics into a near art form. Following copulation, they do not sit on their mates or even spend time guarding them. They don't have to. They simply seal the female's vagina with a plug of cement.

Among thorny-headed worms, males also seek out and force themselves on rival males, cementing shut their rivals' sperm tubes. More devious still, in some other worm species, a male will inject sperm directly into his rival's body, whereupon the sperm moves to the victim's testicles. After his victim recovers and subsequently mates with a female, he impregnates her with the aggressor's sperm.

After sex, the sperm wars continue. Before depositing his own sperm, a male damselfly uses his shovel-shaped penis to scoop out the female's reproductive tract — removing any sperm from previous paramours. Males in other species inject powerful spermicides — or in the case of some sharks, a simple stream of seawater — before mating, destroying, and washing away their competitors' sperm.

Toxins in fruit fly semen, designed to destroy sperm from competing males, also decrease female lifespan by 10%. Nasty and brutish, males battling in these sperm wars care little about females. If nothing else, though, they give us reason for hope and celebration at the relatively benign nature of the battle of the sexes in our own species.

⌐

How can we strengthen our romantic relationships? Humans commonly profess a desire to be monogamous, and the good news is that so many succeed. For all the benefits of marriage, the promise to be faithful necessitates self-restraint. Each partner gives up some freedom in return for a satisfying relationship built on trust. Infidelity is a wanton disregard for the terms of this marriage "contract." To avoid strife, we need to focus on making our marriage a sweet deal.

Step 1 in a mutually satisfying deal is to *do* what we've promised to do. Choosing to spend time with someone is a strong signal of desire and commitment. A husband who provides a constant flow of gifts cements the marriage vows. Women should also give gifts to their husbands, but the best gift may be enthusiastic sex.

In the early days of courtship, what characterizes new lovers' behavior? They long to spend time with each other: no effort is too great to orchestrate a rendezvous, no activity too silly. They give gifts: flowers for no reason or a book they imagine will have meaning for the other.

New lovers listen to each other and spend entire weekends in bed. Is it any wonder these people are so desirable to each other? Don't dismiss such behavior as fanciful and impractical. These couples are crafting the terms of a mutually satisfying deal, and we would do well to emulate them. We shouldn't let our relationships outgrow romance.

Step 2 is *not* to do what we've promised not to do. The surest way to jeopardize a relationship is to renege on the fundamental deal. Women are promising paternity to the male, so any sex with other men will rock a relationship to its foundation. Men are promising to devote their energy to the relationship, so the biggest betrayal is a substantial diversion of time and money to other women.

The temptations we all face are deeply ingrained in the genes of our hearts and minds, and both parties should take steps to fight these mean genes. When we have a close friend of the opposite sex, recognize that this is inherently threatening to our partner. We should include both our friend and our lover in joint activities and share social information freely. For example, listen to phone messages together. Building secret relationships increases both suspicions and temptations.

Finally, we should continue improving ourselves as a way to improve our relationships. Imagine your partner at a cocktail party, enchanted by the witty, smart, vibrant person across the plate of hors d'oeuvres. Is that charmer you? At some point it was. And as long as we remain interesting dynamos, there will be no conflict between monogamy and our infidelity-promoting mean genes.

FAMILY, FRIENDS, **AND FOES**

Family The ties that bind

We love our crazy families. Terry's big sister Sue is the oldest of his siblings. She basked in the glow of her parents' undivided attention in the two years before Burnham baby number two, Jane, came along. Almost immediately Sue resented the attention heaped on baby Jane. This resentment built and built until one day she "kidnapped" Jane in her baby carriage, abandoning both baby and buggy more than half a mile away.

Though cruel (and somewhat creative, for a two-year-old), Sue's actions aren't really that unusual, are they? Family bonds are emotional and tight. The relationships we have with our parents and siblings, in particular, are among the most important in our lives. With so much at stake, it's little wonder that affection and exasperation can coexist so easily.

Take Thanksgiving — or any holiday, for that matter. Many of us see our families only a few times a year, and it's generally something we look forward to. We put time and energy into buying the perfect gifts; we anticipate long, deep conversations with a brother we haven't seen in ages or a favorite aunt. But it doesn't take long for sweet expectations to turn sour. After about five hours at home with crazy uncles, drunk grandmothers, insane cousins, and sulking siblings, we remember why we were so eager to go off to college.

The television networks know how to capitalize on this love-hate relationship. From *Leave It to Beaver* and *The Partridge Family* to *Family Ties* and *Home Improvement*, we get great laughs watching our own sagas played out on the little screen.

Families and kinship are central to every human culture. In the early 1960s, Irven DeVore, a friend and mentor to both of us, went to live in the Kalahari Desert with the !Kung San. Soon after he arrived he was ceremoniously given a name, !Nashe !Na, and assigned a mother (who went on and on about how difficult it had been to give birth to such a big man).

Irv's adoption was more than a simple welcoming gesture. The San have strict, if unwritten, rules of behavior — who you share food with, who you travel with — all driven by kinship. When two people come together, even the language they use is driven by kinship. Never, for example, use baudy words with a mother-in-law, but feel free to crack a dirty joke with a sibling of the same sex. Without Irv's incorporation into the San kinship system, he would have been adrift.

Among the Yanomamö, too, kinship is integral to nearly every aspect of life, even choosing a spouse. Every young Yanomamö is urged to choose their partner from a group of individuals known as their *suaböya*. Literally translated, *suaböya* means "marriageable partners"; they are an individual's cross-cousins. In English, there are no special words that refer specifically to cross-cousins, but we still have them. A person's cross-cousin is either Dad's sister's children or Mom's brother's children.

Although the word *suaböya* is particular to the Yanomamö, many cultures favor such unions. Charles Darwin, for example, married his cousin Emma. His older sister, Caroline, also married a cousin, who turned out to be Emma's brother Josiah.

Cross-cousins turn out to make excellent spouses for several reasons. There are genetic advantages to producing children with someone close enough to share some common ancestry yet distant enough to avoid most of the problems associated with inbreeding. They also strengthen alliances, for instance, between families that are already close due to their common ancestors.

As important as marriage is, for humans the pinnacle of unconditional love may be that between a mother and her child. Still, even the most devoted nurturers among us must be impressed by the Australian social spider. Soon after giving birth to about a hundred hungry spiderlings, Mom's body literally liquefies into a pile of mushy flesh. The babies then munch on the flesh so they can start their lives with full bellies.

Why has evolution produced a dissolving mom? Isn't it enough to drive the kids to soccer practice and make sure they have brushed their teeth? Well, genes are clever and cold, and they build organisms that succeed through various means. Their idea of success is limited to a single goal: increase market share in the next generation.

All things being equal, the genes living in the mom would prefer to live for another day. All things are not equal, however. The death of the mom, although costly, is more than compensated for by the head start it gives the many babies, each carrying copies of Mom's genes.

Genes have built parents that "selflessly" give everything for their offspring, but this isn't the only way to show family devotion. Organisms share genes with cousins, aunts, siblings, uncles, and more, too. We ought to see animals that also make sacrifices for these relatives, and we do — all the time.

⤳

We go to bat for our relatives. Among turkey-like Tasmanian hens, many females live with just one male, but there is also a large number of polygamous females that keep a pair of males on hand instead. This female literally rules the roost: she allows both males to mate with her and requires both to provide food for her babies. This is a pretty sweet arrangement for the females. One study found that hens with two males have an average of 9.6 babies, while those with just one guy managed only 6.6 babies.

Do the males protest? To the contrary. Beyond merely tolerating it, they seem completely unruffled by their *Three's Company* arrangement. Even some of the biggest males welcome small competitors into their nest with little objection. Given that a male with a live-in competitor has 4.8 babies on average (half of the 9.6 that are born), why wouldn't he shoo away the second male and reap the full 6.6 offspring?

Brotherly love is the answer. Males sharing a nuptial nest are usually brothers. And because both males' genes come from the same parents, each shares half of the other's genes. So the open-minded male gets 4.8 babies of his own, plus genetic credit for half of his brother's babies, bringing his total for the season to 7.2. Suddenly the female's exploitation doesn't feel quite so bad. It's just a sly way for brothers to look out for each other.

In a handful of human societies as well, women simultaneously have more than one husband. As with the Tasmanian hens, these marriages only work when the co-husbands are brothers. In one Tibetan society where this form of marriage is common, one of the brothers must often travel on long trips to sell harvested crops. Naturally, this is when the arrangement works best. When the two men are both home, they tend to compete for sex with their wife, although the close genetic ties reduce the tension somewhat.

Tension of a different sort rises high in squirrel families under attack from eagles. Fortunately, these wily creatures have an effective neighborhood watch program. They scream their heads off when they spot a predator on the horizon, telling everyone to run for cover. All that yelling draws attention to the caller, making it more likely that the screamer will be eaten. One out of ten times that a predator appears, a squirrel is killed. And half the squirrels that get eaten are the loudmouths sounding the alarm.

Who would risk calling attention to themselves in order to identify the threat? Why not lay low and live? More than ninety percent of the warning calls are made by females. But it's not exactly out of the goodness of their hearts. Once again, it's about family values.

Female squirrels don't move far from home when they grow up. Males, on the other hand, move far away and continue moving each year. As a result, males never live near their parents or siblings. Ever the drifters, these loners rarely live near their own offspring. Without any relatives to protect, they have little genetic reason to stick their necks out by making risky warnings.

Females, on the other hand, surrounded by members of their extended family, have tremendous incentive for taking the risk. And their genes' shrewdness goes a step further. Almost as if they're maintaining mental lists of their kin, the more relatives a female has in the neighborhood, the more likely she is to sing the "we are family" alarm call.

From a gene's perspective, even a minor nuisance like death needn't be an impediment to looking out for your relatives. When people write wills, they leave the bulk of their money to relatives. Moreover, when people die without wills, governments tend to divvy up assets in a way that mirrors genetic interests.

Kinship drives more than just the flow of money. In 1997, for example, over four thousand living Americans donated kidneys. One of these donors actually gave a kidney to a non-relative. Her act was so rare that she received a flood of media attention. People even stopped and congratulated her in public. Each year, over two thousand Americans die waiting for kidneys and anyone with two kidneys could save a life. Virtually no one makes this sacrifice for non-relatives, though.

Animals, too, go out of their way to look after the best interests of their extended families. For example, the hives of sweat bees resemble Studio 54 in its prime. Bees swarm around, seeking permission to enter from bulky bouncer bees. Getting past the velvet rope, however, doesn't depend on attractiveness or exotic dress. This scene is more like a family reunion than anything else, since the guard bees grant access to close kin and exclude non-relatives with uncanny accuracy.

Tadpoles are similarly nepotistic, preferring to "school" with their siblings. Even when eggs are separated before birth and dumped into a big pond, the tadpoles track down and hang out predominantly with their brothers and sisters.

How do these animals recognize each other? In both tadpoles and sweat bees, olfactory clues make it possible for individuals to assess everyone's genetic similarity. If she smells like family, roll out the red carpet. Humans can be somewhat less sophisticated in distinguishing kin. Jay, for instance, can still hear his mother's voice reminding him each Thanksgiving: "Be nice to Jeffrey, Tammy, Julie, and Karen. They are your cousins."

~

Blood in the family. Here's a paradox: if we love our families so much, why is there so much domestic violence? The *Mean Genes* perspective helps to clarify this apparent contradiction. Approximately one-quarter of all U.S. murders occur within families. Men kill their wives and women kill their husbands. (Perhaps surprisingly, about 30% of murdered spouses are men.) Men kill their stepchildren, too.

Notice the pattern? In the vast majority of family killings, the victim and murderer do not share any genes. One study looked at all 98 family murders in Detroit in 1972. Seventy-six of the victims had no shared genes with the killer. Twenty-two were recorded as children, parents, or "other relatives," and these include some stepfathers.

Studies of many societies, including those of thirteenth-century England and modern Canada, also report very low rates of murder by genetic relatives. When two people gang up to kill someone, however, the perpetrators are frequently blood relatives.

Among the Yanomamö, too, we find confirmation that blood is thicker than water. Whenever villages get too big, tensions rise and tempers flare. Eventually the groups break up and go their separate ways. And who goes with whom?

Here's where things get complicated. It's not always easy to figure out who will stick together because the Yanomamö have a large "fictive kin" network. In this system, people refer to others using kinship terms, much as many American children might call a family friend "uncle" or members of a sorority consider themselves "sisters." Still, the villages split in a way that closely reflects only the actual genetic relatedness. Suddenly they're all able to distinguish between real and fictive brothers.

Child abuse, too, follows a similar pattern, shedding light on the origins of stories such as Cinderella. Researchers reviewed all 87,789 cases of child maltreatment in the U.S. in 1976. They found that children were one hundred times more likely to be killed by a stepparent than by a genetic parent.

$\backsim$

Can we learn to treat everyone like family? So we humans, like other animals, are built to be especially nice to our relatives. Indeed, social dreamers have conjured up worlds in which humans treat everyone as family. Plato, normally a keen observer of people, felt that in an ideal state the rulers should be barred from holding private property. Even after the collapse of the Soviet Union, utopian dreams live. Can we learn to be equally nice to strangers?

Just a few years ago, Portland, Oregon, bought eight hundred bicycles for common use, arguing that this socialistic solution was infinitely more efficient than having thousands of privately owned bikes sitting in garages mostly unused. The fleet was soon reduced to a handful of poorly maintained wrecks. Some of the bright yellow public bikes were seen being loaded into a pickup truck with out-of-state license plates.

Our problems extend far beyond bike theft. During 1998, there was a violent crime — murder, rape, robbery, or aggravated assault — reported for every 177 people in the United States. There was a property crime for every 25 people. You know several of these victims. Every few years you *are* one of these victims. Fortunately for our national morale, the FBI doesn't bother to track how frequently people merely lie or act selfishly.

Despite these unpleasant facts, we continue to hope that humans can be selfless. There is a definitive answer to this question about human organization, but we have to look in some unusual places.

One fable describes a leadership battle among parts of the human body. The eyes argue for the command position based on the importance of vision for all human endeavors. "Where would we be without my metabolic work?" counters the liver, which then reminds the others that it also processes alcohol.

The brain, somewhat haughtily, claims the mantle due to its superior IQ.

Meanwhile, the intestine goes on strike, refusing to process waste. As poisons accumulate, the liver chokes, the eyes tear, the brain becomes foggy, and the intestine is crowned king.

This is one area where we are free of conflict — the internal running of our bodies. The fable of the intestine king is humorous precisely because it is false: human eyes, liver, brain, and intestine all work selflessly toward the good of the whole body. In a communistic ideal, our immune system cells throw themselves into lethal combat against invading diseases without battle medals or patriotic speeches.

Why doesn't the liver go on strike to get a better deal? Assume for a moment the view of a gene in the liver. Does it gain if the liver wins additional energy flows from the rest of the body and is granted shorter work hours? No. The liver gene has one and only one road to genetic success: help the body have sex and babies. The genes in the sperm and eggs are derived from exactly the same pool as those in the liver. Genes in all parts of the body further their own interests best by cooperating.

In addition to our bodies, there are many societies where every single individual works for the public good. Unfortunately, everyone in these societies has six legs. Ants and bees make perfect communists; every individual puts the needs of the group first. Police needn't patrol the nest or hive. Ministers need not preach fire and brimstone to keep individuals in line. Their societies resemble finely tuned machines.

Stir up a honeybee nest and you'll get a taste of bee altruism. Bees happily make the ultimate sacrifice for the greater good of the hive. Death occurs when they bury their stinger so deep into human skin that the abdomen is ripped apart. Why do the genes in the suicidal honeybee care so little about their own future?

Individual worker honeybees are sterile; the genes they carry win the Darwinian competition only when their queen mother spits out more offspring. Bees willingly perform any act to help the queen, including dramatic, suicidal stinging death, which protects the hive from honey-seeking intruders. Genetically speaking, the hive is like our body, and the honeybee, like a cell from the immune system. All for one and one for all.

Mud daubers look just like honeybees, but they are much more hesitant to attack people. Practicing their own family values, mud daubers form pairs, have sex, and raise offspring. Unlike our honeybees, each mud dauber is fertile, so individual death means genetic loss. While honeybee genes win through brave sacrifice, mud dauber genes gain through judicious cowardice.

Let's return to the question that Plato grappled with: can selfless human utopias survive? The answer is a disappointing but definitive no. It's a bit ironic, but after hundreds of years of utopian dreams and failed social experiments, scientists found this answer by understanding beehives.

The genetic interests within ant colonies, honeybee hives, and our own bodies are aligned in such a way that there is no conflict. Conflict arises only, and inevitably, between entities with different genes. Ant colonies are communistic within but continuously wage war on their ant neighbors. Similarly, although we truly love our families, when push comes to shove, our genes will always fight to keep our own genetic interests first.

Families are not free of conflict either. Here's a disappointing truth: a mother cannot count on as much devotion from her baby as from her own liver. The mother and child's mutual love is tempered by Darwinian reason. Half of the baby's genes come from Dad. This inevitably drives a wedge between a pregnant mother and the fetus she carries and loves.

Specifically, the mother and fetus disagree about how much food — doled out as nutrients in the blood flow across the placenta — the fetus ought to get. Deep love notwithstanding, there's a point when Mom wants to stop pumping glucose and other treats to the baby within. Why would she withhold the goodies? Because her genes gain if she saves a bit for future fetuses. When setting the support dial, a mother balances her own needs — in particular, those of her future offspring — against those of the fetus.

Now consider the view from the fetus. The genetic calculus is different; future siblings will carry some but not all of the fetus's genes. Consequently, the fetus is not as keen about sacrificing for their benefit, and screams for greater support than is ideal for the mother's genes. Sibling rivalry starts before the sibling is even conceived!

This conflict results in silent strife throughout pregnancy. The fetus pumps out hormones that dilate the mother's blood vessels. This increases the amount of sugar in the blood — and, consequently, the size of the fetus's meals. Mom retaliates by producing insulin, which has exactly the opposite effect. In some mothers this conflict causes diabetes — which disappears as soon as the baby is born — and in all pregnancies it escalates until mom is producing a thousand times the normal amount of insulin.

The mother-child discord doesn't end with birth. A study of all births in the United States between 1983 and 1991 concluded that 2,776 babies were killed by their mothers. It further revealed that the rate of infanticide rose during the study period. We began with mothers who die for their offspring, and now we're discussing mothers killing their babies. What's going on here?

Infanticide, like maternal diabetes, happens because the mother and child have shared but not identical interests. A mother loves her children, but she can have other babies. Mothers don't kill just any baby: there's a genetic calculation at work. The most comprehensive survey of human infanticide, covering dozens of societies, found that mothers kill their own infants when the baby arrives at a time when it can't be supported.

In many cultures, the question of whether to keep a baby is entirely the mother's, and everyone respects her decision. Among the !Kung San, for example, women typically give birth with the help of a close female relative. If Mom brings the child back to the group, it is recognized as a person. If no child returns with the woman, it is assumed that the child was stillborn, regardless of the circumstances.

What have we learned about families? Genes play a central role in fostering cooperation. At one extreme are socialistic collectives, where everyone pulls for the good of the group. For ant colonies, beehives, and our own bodies, Marx would have smiled as the motto "From each according to ability, to each according to need" is played out.

Whenever genetic interests do not align perfectly, however, conflict is inevitable. George Washington's presidential farewell speech focused on foreign policy, concluding that the country has permanent interests but no permanent friends. People face a similar social world, and as we'll see, all human relationships require continual diplomatic maneuvering between conflict and cooperation.

Friends and Foes Keep friends close
and enemies closer

Is conflict between societies inevitable? Antagonism among family members is almost universal, yet nothing causes us to put aside our differences more quickly and band together than a threat from outside the family. An ancient maxim says, "Me against my brothers, me and my brothers against my cousins, me and my brothers and cousins against my clan, me and my clan against the world."

A folk tale tells the following story. A goddess visits a farmer and promises to grant one wish, but with an interesting twist. The farmer would be granted his wish, and each of his neighbors would receive the wish in duplicate. After a moment's reflection, the farmer asked that half his crops be destroyed. This tale reminds us of an unpleasant reality. In a world with finite land, food, and mates, one party can sometimes gain only by taking from another. Pain for a competitor can be good, even at a cost of half of your crops.

Our close genetic relatives, the chimpanzees, have instincts that seem to incorporate this sense of scarcity. On January 7, 1974, researchers at Jane Goodall's Tanzanian site witnessed a new form of chimpanzee aggression. Eight members of one community surprised and killed a young adult male from the neighboring group.

Over the next three years, the aggressors completely wiped out their neighbors and took over their territory. Researchers witnessed a half dozen of these assaults on males, females, and even infants. This behavior was particularly shocking because just a few years earlier the two groups had been one. After what appeared to be an amicable split, the groups became antagonistic and the smaller one was eradicated.

Previously chimpanzees had been known to defend their own territory, but this was the first documented case of an organized territorial invasion that resulted in the death of the victim. Since then, however, observers have seen similar behaviors, termed "lethal raiding," in other wild chimpanzee populations.

In each case there is a common pattern. Groups of chimpanzees, usually males, patrol their territories and opportunistically exploit numerical advantages to enter the adjacent community to attack their neighbors. Chimpanzees going through the forest are generally very noisy, because of both their movements and vocalizations. In contrast, those on border patrols and raids are eerily quiet.

Human societies frequently display similar territoriality. Why are some cultures intent on taking over their neighbors while others are content to stay at home? Some interesting clues come from examining the Polynesian region of the Pacific. Between 1200 B.C. and A.D. 1000, many of these islands were populated by people of the same genetic and cultural heritage. Some became warlike and others did not. Why?

In a word: agriculture. Many islands were too cold to support crops, and the inhabitants survived by hunting and gathering. This is a relatively tough way to get calories and, consequently, the populations remained small. These cultures had loose political structures, no armies, and peaceful outlooks.

To the south, life grew steadily easier. The inhabitants of these islands were able to cultivate crops in the warmer territory. Their more dependable and plentiful food supply led to bigger families. But their populations soon became swollen and, running out of room to grow, they became warlike. These cultures stored food in great quantities, developed military skills, and fought one another.

The lesson is clear: along with high population density comes competition for resources, and with competition comes conflict. Across a wide variety of cultures we see a similar pattern. The !Kung San, for example, subsist at low population densities in the Kalahari Desert and have the same peaceful organization as that of the Polynesian foragers.

Are humans thus naturally warlike? Not necessarily. It is true that we have a long history of territoriality and intergroup aggression that presumably dates from our chimpanzee-like ancestors. It is only in certain competitive environments, though, that conflict is manifest through warfare. Unfortunately, certain innate human traits tend to foster these environments. For one, we tend to form group affiliations very quickly. Indeed, participants in psychology experiments pick up group identities almost immediately.

In one study, the people were divided into groups along arbitrary lines — let's call them "the blues" and "the reds." They spoke within their groups for a few minutes, then played cooperative games for cash prizes. Although the groupings were arbitrary and the amount of money they won was not shared even within the group, the reds were nicer to other reds and blues nicer to other blues.

Outside the laboratory, our affiliations can endure for years. Paul, a friend of Terry's, recently found himself alone shouting at his TV, "DIE PARCELLS, DIE." Bill Parcells, a legendary football coach, had recently deserted Paul's New England Patriots to coach the rival New York Jets. Normally mild-mannered, Paul worked himself into a frenzy watching his team battle the hated Jets. His enmity was fanned by seeing the traitorous Parcells strutting his stuff on the opposing sideline.

Professional sports are a relatively harmless outlet for our taste for aggression, victory, and crushing rivals. In a variety of competitive settings, victory causes sportsmen to have higher levels of testosterone than their vanquished foes. Pinning an opponent in wrestling, for example, is accompanied by a physiologically induced winning glow.

Recent studies reveal that sports fans share viscerally in this biological victory dance. Fans of winning sports teams have higher testosterone levels than those of losing teams. We feel powerful emotions when we watch sports because we are having hormonal surges. Exulting as the Patriots crush the Jets can produce a victory feeling as profound as that after a battle, with none of the casualties.

When Roman armies won battles they returned home to outlandish ceremonies. Similar spectacles accompany sporting victories. When a team wins the World Cup in soccer, the whole country goes wild. When the Philadelphia Phillies won the 1980

World Series, five hundred thousand people turned out for the victory parade in a city of 1.7 million. This may be a higher percentage than at the enormous victory celebrations marking the end of World War II.

If humans bond over superficial or arbitrary groups — reds, blues, Patriots, Jets — what about over race?

～

Race and biology. In a fraction of a second we are aware of someone's race (along with other prominent characteristics such as gender, size, and age). Because history is filled with so much racial tragedy and oppression, however, many of us feel awkward discussing it or even mentioning race.

One humorous outcome is the difficulty in following televised boxing matches. When the boxers are of different races, commentators say, "Lewis, in the red shorts, is aggressively pursuing Jones, in the blue shorts." Viewers must constantly remind themselves that the black man is wearing blue shorts and the white man, red.

Traveling through Kenya a few years ago, Terry noticed something interesting. Each time the little safari group stopped along the road, the Kenyan tour guide would immediately ask any other Africans they encountered, "What tribe are you?" Kenya's people come from more than a dozen different tribes, and historically, members of one tribe could be easily identified by their traditional clothing, locations, and body ornaments.

In modern Kenya, however, many of these visible tribal signs have disappeared and a large percentage of the population dresses in Western clothing. In their T-shirts and blue jeans, many tribes now look so similar that members cannot identify one another. Nevertheless, tribal affiliation remains a central feature of the interaction between Kenyans.

When two Americans meet these days, a similar ritual involves college affiliation, with loyalties immediately laid bare. Don't expect a "Wolverine" from the University of Michigan to warmly embrace one of the "Bruins" from UCLA.

Our practically compulsive need to categorize extends to trivial features. Humans pay inordinate attention, for instance, to subtle differences in clothing, accents, and myriad other cues. In the fourth grade, Jay was mortified when his mother inadvertently bought him sneakers that had four stripes. This was an unmitigated social disaster. All of the "cool" kids wore Adidas, with their trademark three stripes. Jay was able to save face only by using scissors to carefully remove one of the stripes.

If we can quickly discern the difference between European and American jeans, it should come as no surprise that we notice race and ethnicity. Psychological studies show, in fact, that people classify race unconsciously and almost instantly. In one set of experiments, for example, racial information visible on a screen for only an instant altered the subjects' reaction time.

In the TV show *All in the Family*, the lead character was Archie Bunker, a stereotypical white racist. Although the show was considered a comedy, one of its running themes was a serious exploration of troubled American race relations. In one show, Archie refused to donate blood because he was afraid his fluids would be mixed with those of other races. Focusing only on surface differences, Archie failed to see the underlying but over-whelming similarity.

Are blacks and whites genetically different? The answer is obviously yes; black skin genes are different from white skin genes. Moreover, the prevalence of some genetic diseases varies by race and ethnicity. Ashkenazi Jews, for example, are more likely than other people to suffer from the genetic disease Tay-Sachs. Similarly, sickle cell anemia is relatively common among Africans and Southeast Asians because the genes that cause it also improve a person's resistance to malaria, a disease prevalent in Africa and Southeast Asia.

Push this analysis a little bit further, however, and our simple categorization of races disappears. The genetic resistance to malaria, and the consequent higher risk of sickle cell anemia, is shared by both Africans from the south side of the Mediterranean and Europeans from the northern side.

People in both groups were bitten by malaria-infected mosquitoes; consequently, both evolved the same genetic defense. On the other hand, Africans from the southernmost part of the continent have no greater risk of sickle cell anemia than do the Japanese because malaria is similarly rare in both of their homelands. So for this trait, the southern Africans more closely resemble the Japanese than they do the North Africans.

It's misleading — and can be dangerous — to draw conclusions from the fact that whites and blacks are genetically different in a few genes that happen to be really obvious.

The assessment of race from a genetic perspective is fraught with problems at every level. First, how do we even decide who is white and who is black (or who is Basque)? Walking down Main Street, USA, this may seem easy. But if you travel from tropical Africa up through Egypt and into the Middle East, it is impossible to discern where one race ends and the next begins.

Race is about as useful a distinction, genetically, as height. At the extremes it's possible to classify people as tall or short. But is the man who is five-foot-nine short or tall? What about five-ten? In a world filled in which people fall at every point along the continuum, grouping becomes arbitrary. Similarity in height between two people is a terrible proxy for genetic similarity.

Nevertheless, there are some genetic differences between races — at least with respect to skin color and hair form — and our instincts notice those differences. But more than 100,000 genes make up an individual, and for the majority of them, every single person on the planet is identical. The ability to swap blood and other organs between individuals with different skin colors reveals this overwhelming similarity.

Using advanced DNA technology, measures of genetic variation confirm that human races are trivially different from one another. For that one-quarter of all our genes in which there is some variability, there is little rhyme or reason to how this variation is divvied up from one person to the next. Africans have huge variation in blood type: some are type O, some AB, others A or B. But the same goes for Asians and Turks, Russians and Spaniards.

Beyond visible traits such as skin color, Europeans also have tremendous variation in a bunch of naturally produced proteins with tongue-twisting names, such as 6-phosphogluconate dehydrogenase and adenylate kinase. Some Europeans produce lots of these compounds, others almost none. But again, the same is true of the Inuit and Navajo.

If an asteroid strikes the earth and kills everyone except for the people now living in Africa, 93% of all the human genetic diversity would still be present. On average, we'd have slightly darker skin, but we'd have pretty much all of the same genes that we currently have. Put simply: among humans, race gives few clues to which genes an individual carries.

This contrasts sharply with race differences among some of our closely related species. Compared to the other great apes, humans turn out to be an unusually homogeneous group. As a result of a nearly complete lack of migration and interbreeding for millions of years, lowland gorillas, for example, are very different from Dian Fossey's gorillas in the (mountain) mist.

Unlike gorillas, any two humans — even two with different skin color — are significantly more similar genetically than a lowland gorilla and a mountain gorilla. No human race is completely isolated from all of the others, so human racial groups have not developed genetic differences as profound as those in gorillas. Further, with our extensive movements around the globe and frequent intermarriage between groups, it is likely that the existing genetic differences between races will shrink.

When we talk about racial difference, we are concerned with genetic differences between groups of people. Another measure comes from computing the total genetic diversity within groups. Here, too, humans show a remarkable genetic similarity from individual to individual as compared to other apes. In fact, a re-

cent study found more genetic diversity in just fifty-five chimpanzees from one community than is found in all six billion
humans.

We are left to deal with a strange combination of features that
causes trouble. Humans have a long history of conflict between
groups, we are very prone to forming coalitions even along arbitrary "red" and "blue" lines, and we have surface differences
that are immediately obvious. Yet underneath, we are practically clones of one another. So race is not a figment of our
imagination, but it is largely a figment of our perceptions.

~

Bitter enemies sometimes cooperate. Amid human conflict, again
and again we find that cooperation creeps in. Even on the
bloody battlefields of World War I, individual units facing each
other across the trenches showed an unexpected ability to initiate mini-truces spontaneously without formal agreements
or verbal exchanges. How did these adversaries find common
ground?

This cooperation was always precarious, and the peace agreements would periodically devolve into fighting. But time and
time again, peace broke out. By analyzing reports from both
sides, researchers have identified some of the critical conditions
that made peace possible. First, one side had to make a gesture.
In the middle of the killing, there had to be some respite. In the
trenches, one side might begin purposely misdirecting artillery
fire toward an empty position, for instance.

Alternatively, one side might open fire at a standard time. The
British in one sector, for example, might begin a daily artillery

bombardment at precisely 1:00 P.M. After several days, the Germans would begin lounging outside their trenches, enjoying the sunlight until 12:45 P.M. They would then head into their deep fortifications as the British bombs dropped harmlessly around them.

Another essential feature to maintaining these truces was that punishment, albeit limited, had to be meted out in response to any deviations. If one side broke the informal rules and actually tried to kill someone or fired a barrage before the appointed hour, the other side would retaliate.

One regiment even codified this into a rule of thumb: never shoot first, but when fired on, give back exactly twice what had been given. Inherent in this two-for-one punishment is the idea that once it takes place, there is real forgiveness. The punishment evened the score and allowed the sides to return to their earlier truce.

The last essential condition for cooperation in the trenches was the clear identification of the individuals on the opposing side and an ongoing relationship with those individuals. Cooperation takes time to develop, and the fear of punishment can succeed as a deterrent only if the offender will be punished.

‿

Understanding conflict can improve our relationships. Does cooperation during wartime have any relevance to our lives in more peaceful settings? Absolutely. The keys to good relationships in peacetime are exactly the same as in warfare. While we may all dream of unconditional friendship, we are just as motivated by self-interest as the troops.

Some other animals have also learned to cooperate for similarly selfish reasons. Take vampire bats, for instance. As their name suggests, they live by sucking blood from other animals (though rarely from humans). These flying parasites cooperate by sharing food. If a bat returns from an unsuccessful hunt with an empty belly, it solicits — and often receives — a regurgitated meal from a roost mate.

The value of swapping blood is dramatic; the bats are never far from starvation — just two nights without a meal means death. Thus, every bat can gain from a system of swapping excess blood on good days for starvation insurance on bad days.

Most other animals could make similar mutually advantageous swaps. On a rainy day, I borrow from you with a promise to repay when my own sun is shining. We can both come out ahead if each gift benefits the recipient more than it costs the donor. For this reason, we might expect all animals to be altruistic in this sense — that is, the selfish sense of giving in times of plenty that is likely to be reciprocated in time of need.

Surprisingly, though, across the animal kingdom we find that vampire bats are almost unique in their friendly disposition toward non-relatives. Many animals cooperate in other ways, but they do not engage in reciprocal exchanges over time. For example, one cooperative behavior is shared territorial defense, but it brings immediate advantages to both parties.

So the reciprocal granting of favors between non-relatives rarely happens. Part of the reason is that many animals, including our close relatives the orangutans, are generally hostile toward other members of their species except mates and offspring. Humans, however, are extremely social. We like to be with others, whether over the carcass of a gazelle, a sushi dinner, or the water cooler.

Furthermore, unlike members of most species, we easily form cooperative arrangements requiring future repayment. Imagine that you forget to bring your wallet to work one day and need to borrow money to buy lunch. Will it be difficult for you to eat? The answer for almost any human is that it is easy to get a favor that will be repaid. Such easy credit between individuals who are not family is never seen in non-humans. Why do people cooperate so successfully?

Simply put, because we have everything it takes to avoid being duped. Cooperation is rare because it's dangerous. Accepting the short end of the stick is a sure road to extinction — it may be better to not cooperate at all. As a demonstration of our innate skills, answer this: Which of your friends is a little cheap when it comes to paying the dinner tab at restaurants? Did some of the guests at your wedding fail to send a present? Have you sent a holiday card to people who didn't send one back?

We may claim not to be so petty, but be honest: could you answer the questions, identifying your cheapskate friends? To avoid exploitation, we all keep detailed mental lists of favors owed as well as benefits received and given. Eventually, inevitably, we end relationships with people who do not reciprocate.

The instinct that prods a human to share food with an unrelated Stone Age neighbor, to loan money to a twentieth-century neighbor, or just drive a friend to the airport appears to be refreshing altruism. It isn't. Just as we put money in the bank for a rainy day (or put fat on our thighs for a no-grainy day), we buffer ourselves from the world's uncertainties by storing good-will in our neighbors.

Niceness and cooperation are simply subtle forms of selfish-ness. Cynical? Perhaps, but as we will see, human brains are built to monitor cooperative relationships. Consider two differ-ent sources of kindness. The first is the fairy book brand, in which people give of themselves simply to create happiness; they put money in anonymous collection boxes and never care if they are repaid.

The second is the selfish version, in which accounts are main-tained and deviations from norms are noted and punished. Un-doubtedly modern human behavior combines the two, but let's examine the mechanisms we all have that ensure that our acts of "altruism" ultimately work to our benefit.

The black hamlet fish of the Caribbean has an interesting coop-erative problem. Each fish has both male and female sexual or-gans, and having progeny requires finding a partner willing to play the other sex. The wrinkle is that eggs are big and expen-sive compared to sperm, so each fish in a pair would rather play the male.

The egg-laying fish is at risk because the sperm releaser gains 50% of the genetic benefits but expends less than 50% of the energy necessary to produce offspring. The only equitable solu-tion — and exactly what we see in nature — is an alternation of roles: "Lyndsey" lays a few eggs that are fertilized by "Jamie," who then must take a turn as egg producer. If Jamie does not re-ciprocate by laying some costly eggs, then Lyndsey leaves.

Humans have a similar taste for fairness. In laboratory settings, people are willing to pay to enforce fair outcomes in the "ultimatum" game. In it, two people squabble over a pot of money. One person gets to propose a split. The other has just two choices, accept or reject the proposal. The first offer is an ultimatum because it is final: either the pot is divided as proposed or it's destroyed.

For example, let's say the pot is $100 and the proposal is $90–$10. The person getting the short end of the stick has two choices. Take $10 and see the other person walk off with $90. Or reject the $10, in which case neither party gets any cash. What do you think happens when this game is played for real money, often high stakes?

To help you answer, imagine that the "pie" is not money but the evolutionary benefit from cooperation. To give a human example, perhaps one of our ancestors could hunt alone and capture a small amount of food, but joint hunting would capture large, calorie-filled beasts. Human cooperators, like the sharing vampire bats, could outbreed loners. This fact underlies human sociability.

From archaeological deposits, we know that our hunter-gatherer ancestors lived in groups of a few hundred people at most. Their success depended on joint efforts against predators and prey; being nice paid off when the prospect of hunting alone or sleeping outside the camp meant death. The loners died, so we are descended from those who could work well with others.

While cooperators have evolutionary advantages over loners, however, they are at risk if they get the short end of the stick repeatedly. If a master hunter continually did all the work for the rest of the group, he would eventually lose the evolutionary race to his more cunning partners. Black hamlet fish know to avoid exploiters, and so do people.

Returning to the ultimatum game, would you accept $10 if your partner planned to keep $90? How about $1 and $99? Subjects who play these games for real cash routinely smash unevenly divided pies, choosing nothing over inequality. This taste for fairness holds true even as the stakes are increased to many months' worth of salary in poorer countries. Isn't it pure lunacy to turn down free cash?

Perhaps. Nonetheless, most of us feel compelled to punish brazen selfishness, even if we must forfeit free money. It feels less irrational if we think of the money we destroy as a down payment for the future. If we reject someone's stingy offer, we're more likely to be treated fairly in future ventures.

A few years ago, our good friend Patricia did an incredibly generous thing. Knowing that her friend Katherine was in dire financial straits, Patricia loaned her $10,000. Katherine was overwhelmed. Did Patricia want a contract of some form? No, she replied, "I trust you." What happened? The gift ruined the friendship. Even though Katherine is now married to a millionaire, she never repaid the loan. Not surprisingly, the two no longer speak.

Our relationships rest on a selfish foundation, so too large a gift can destroy the balance of a friendship. Our instincts make us constantly weigh the benefits of continuing relationships with their costs. For Katharine, continuing her friendship with Patricia just wasn't worth $10,000.

Moral people honor their commitments even when it does not pay. One of the hallmarks of humans is our ability to override our impulses that often push us toward selfishness. Still, there is no reason to put people in situations where they may be better off reneging. A simple one-paragraph letter from Katherine acknowledging the debt would have held up in court. If Patricia had let Katherine give her such a letter at the outset, the two would probably still be friends.

One of the keys to effective relationships is to maintain a balance of favors. If one person's debt becomes too large, he is likely to end the relationship and skip out on the debt. The black hamlet fish courtship alternates roles to achieve a balanced, cooperative outcome. No fish ever owes another more than one batch of eggs.

In some bird species, the male builds a big nest during courtship and the female reciprocates by laying eggs, followed by their joint feeding of the chicks. Studies show that males and females do occasionally desert their partners, and as expected, they desert at the times predicted by their investment. So males never leave in the short period of time between building the nest and having sex with the female. Rather, they leave after fertilizing the eggs and before parenting begins.

Citicorp and other big banks would have done well to learn this lesson when they loaned money to Latin America during the 1970s. Wise debtors are always weighing the cost of repayment with the cost of default. In the Latin American case, there came a point when ruining the relationships with lenders was less costly to some governments than sending cash to the United States. The result was that most Latin American countries delayed debt payments, or defaulted, costing the banks billions of dollars.

Imbalance ruins relationships, and the riskiest time for partners in trade is always after someone has just made a big investment and it is time to switch roles.

~

Managing our relationships. As we've seen, cooperation is a delicate business. Maintaining friendly relations requires more than balance, however. Consider that in their quest for alpha-status, chimpanzees, even "friends," test one another's physical strength regularly. Even a brief illness cannot remain hidden from such tests, and weak alphas are discovered and dethroned.

Similarly, one side in a cooperative deal may deviate from the expected nice behavior by accident, but such a deviation also probes the strength of the other side. In these cases, perhaps paradoxically, punishment is required in a functioning relationship.

When Don Corleone is nearly assassinated in *The Godfather,* relations between the Corleones and the other families reach a boiling point. The resumption of peace requires retaliation, and reconciliation can occur only after the Corleones kill their enemy's son in cold blood. Cooperation is built on mutual strength; had the Corleones extended the hand of friendship without punishment, it would have been shot, not shaken.

Our punishing natures require accountability. Wrongs must be redressed, but it is important to strike the guilty parties. Fortunately, we have some machinery to help us with this. For example, are you good at remembering faces? What about names? For almost everyone, faces are easy and names are hard, because a substantial chunk of the human brain is devoted to face recognition.

People who develop tumors in this part of the brain have a disease called prosopagnosia; they cannot recognize anyone, including their spouses or even photographs of themselves. Those of us with normally functioning brains can easily remember people, and it is particularly useful to keep track of those we have helped and those who have cheated us. Vampire bats also can remember — and punish if necessary — approximately a hundred different individuals.

Why is face memory so important? Evolution has built us to solve problems that occurred frequently in our history. The ability to remember faces — especially of those who may have wronged us — has long been important in a world in which only a few saints are wandering around bestowing favors but cheaters are common. As a result, humans have a finely tuned, instinctual system for detecting cheaters.

Recall the final feature of the trench warfare cooperation. The units needed a reason to stay friends, in the form of future interactions. We find the same theme in our lives.

Have you ever informed an employer that you were leaving only to find frost immediately descend? What seemed to be a friendly relationship suddenly turns sour, and you wonder why you didn't just take your last paycheck and leave a few hours later. Friendship often depends as much on future prospects for mutual benefit as it does on personality. As the expected length of a relationship shrinks, so too does the cooperation.

So cooperation needs care and feeding and a future. In the harsh, uncertain world of our ancestors, a person with such relationships had a real advantage. Not surprisingly, then, modern humans derive deep pleasure from reciprocal relationships, even to the point of some apparently silly customs.

In *The Treasure of the Sierra Madre*, Humphrey Bogart is part of a doomed mining party returning to town to get medical care for an injured man. They meet some locals, and before discussing medical help the two groups exchange gifts. Bogart notices a strange fact, "We give 'em our tobacco, they give us theirs. I don't get it." Human cooperation is so tied to reciprocal exchange that even apparently senseless gifts and tiny gestures of good faith can play an important role in building relationships.

Vervet monkeys practice a limited form of cooperation by forming alliances during fights. In a form of mutual tobacco sharing, they maintain alliances by grooming each other. Some of the grooming involves removing parasites, but most of it has absolutely no effect on health. Nevertheless, grooming increases the chance that the recipient will come to the aid of the groomer in subsequent dangerous battles.

People in some Pacific islands participate in the "Kula Ring." The islanders form small, warlike groups and exchange gifts with certain neighboring groups. As with the vervet monkeys, the flow of gifts predicts future alliances, this time in wars.

You could interpret these alliances as simple protection of trading partners, even if the trade is called gift exchange. The interesting twist of the Kula Ring is that the gifts are completely useless — necklaces that move from group to group and are never worn. And you wonder why your insurance agent sends you a birthday card every year.

Taken together, these facts and anecdotes expose a selfish understructure to our friendships, gifts, and cooperative ventures.

The brain is an extremely pricey organ, consuming 20% of our energy resources, although it is only 2% of body weight. Much of this precious territory is used to keep track of gift flows, to store faces, and to detect cheaters.

We estimate the durability of relationships and are nicer to those with whom we have a future. To regulate and earn respect, we punish our enemies and even our loved ones for deviations from friendly behavior.

Why is gossip irresistible? One of the most entertaining aspects of our selfish, relationship-controlling instincts is our taste for gossip. Humans are unique because our language allows us to share information rapidly and we love to chatter.

Take the following quiz: What do you know about Julia Roberts's love life? Who is she dating and whom has she married? Now answer this: Will you ever meet Julia Roberts? An amazing number of people who will never, ever even see her, know that she was married to Lyle Lovett, then got divorced, and is now dating an attractive actor named Benjamin.

Gossip filled with juicy tidbits is a human universal, and the amount of time spent nattering away does not go down even if there are just a few people about whom to gossip. Many !Kung San groups contain fewer than a dozen people and you might think they'd get bored talking about one another. They don't, and instead spend hours each day dissecting and transferring such information.

Gossip has a function. We share useful information with our allies about food sources, bargain prices, sickness among our rivals, and sexual opportunities. We also use words to grease coalitions against our enemies, spreading harmful and malicious stories. All of this is very helpful in gaining a leg up in the struggle to survive, prosper, and mate.

But our love of social information has run amok. We devour gossip not only about Julia Roberts and other people we will never meet, but also about fake people. TV soap operas are fictional stories filled with precisely the sort of behavior that dominates the discussions around !Kung fires. Did you know that Kimmy is having sex with John and that she is having his baby? We love this information so much that some magazines' sole purpose is to summarize the social data on this fake world. Why?

Social information about strangers and fictional characters is junk food for our gossipy instincts. For the !Kung and our ancestors, gossip was functional. For us, these tasty nuggets of personal information are "empty calories" — time spent dwelling on people who have nothing to do with us.

～

Etiquette is no trivial matter. In addition to gossip, gift giving is a human universal with a deep history. Gift etiquette can be quite tricky, though, as Emily Post knows and a U.S. official found out on a visit to China. When his host placed a piece of food on his plate, the American ate the gift, then immediately reciprocated with a piece of his own food. The Chinese were aghast.

Commentators explained the mistake as a fundamental cultural difference. American friendship, they explained, is quid pro quo, while Chinese relationships involve less explicit account-

235 FRIENDS AND FOES

ing. Observers failed to note the common feature: people in both cultures give gifts with an eye to repayment; the only difference is the time frame.

When favors are not repaid immediately, the debt can linger for years. In the movie *Gardens of Stone,* a man calls his buddy and asks for a favor: to woo a new woman, he needs the buddy and his wife to come over for dinner. The buddy begs off, and when pushed, says, "Watch my lips, kid. No way, no how." The caller then plays his trump card. He says, "Goody, I saved your life in Vietnam. I'm calling in my marker." The movie cuts to dinner, where we see Goody and his wife.

The most comprehensive anthropological survey of gift giving across all cultures summarized this universal human behavior in one sentence. In all societies, people profess that gifts are selfless and voluntary when, in fact, gifts are selfish and obligatory.

Gift giving can become almost aggressive. The Native American tribe of the Kwakiutl, for example, had a ceremony known as the Potlatch. The host served a feast and distributed gifts with the aim of establishing social dominance. To maintain honor in the face of a competitor's lavish ceremony required an even more expensive feast in return. So nothing is worse than being a guest at the table of a generous host. Even though the Potlatch itself is no longer practiced, similar rituals continue.

On a recent trip to California, Terry spotted some cool art in Jay's office. "Hey, what's that?" he inquired. Learning that it was some of Jay's own work, Terry offered to buy it, even offering $1,000. "It's not for sale at any price" was Jay's response. On his next trip to Boston, Jay carefully packed up the images and presented them to Terry as a gift. They now hang over his couch.

The reciprocal gift has not yet been purchased, but Terry knows it will have to cost far more than $1,000.

❧

Gifts can seal the deal. Last year, Terry decided to renovate his condominium completely. He found an excellent contractor, Patrick, and they agreed on a plan. The contract discussions took their inevitable form, with each side arguing for more advantageous financial terms. After some haggling, Terry and Patrick agreed on a total price in the tens of thousands of dollars.

Along with the first check to start the work, Terry included an expensive bottle of Scotch, suspecting it was Patrick's drink of choice. After fighting ruthlessly for the lowest price, Terry voluntarily gave Patrick an extra $60. Why?

Well, people evolved in a world without contracts, one in which gifts had a central role in mediating relationships. Although we now have a legal system to regulate our interactions, we still have instincts for giving and receiving gifts. Terry's gift was designed to stimulate a positive feeling in Patrick. Imagine how much more effective this $60 was, spent on a gift as opposed to being added to the overall bill. We're not sure of the gift's effect, but the relationship started on a positive note and never soured.

Economists are not generally known for their optimistic views of human nature. Adam Smith's most famous quote, from *The Wealth of Nations,* is: "It is not from the benevolence of the butcher, the brewer, or the baker, that we expect our dinner, but from their regard to their own interest. We address ourselves, not to their humanity but to their self-love, and never talk to them of our necessities but of their advantages." Even hard-hearted economists have data supporting the value of gifts.

In an experimental setting, subjects were paid to simulate bosses and workers. As in all real employment settings, wages were the central feature used to hire workers. After the negotiation was finished, but before any work began, the employers had a chance to give the new workers a financial gift. This unearned bonus did not change the contractual terms. What happened?

Employers who gave the bonus made more money. How? The setting was designed so that workers could choose to work hard or to slack off a bit without being caught (just like most jobs). Production levels went up — along with profits — if workers volunteered for what was essentially unpaid overtime. Workers who received unexpected gifts worked harder — so much harder that employers who gave gifts made more money.

Real companies play this game as well, giving a variety of perks to employees. Many Internet start-ups, for example, tout their fun and generous corporate atmosphere: "Come work for us. We give you food and a gym to work out in." Friday afternoons are filled with company beer bashes. Why are companies generous? It pays. Like the bottle of Scotch, perks manipulate our gift-exchanging instincts. There's nothing wrong with this, and most of us are happier with more than just a quid pro quo.

These instincts have ramifications in many areas. One study looked at people's willingness to be interrupted at a photocopier. Researchers camped out near the copy center and sent in accomplices who asked to use the machine immediately. What happened?

The researchers found a big difference if the interrupter explained the need. "May I make some copies now?" was much less successful than a request along the lines of "May I make some copies now because my boss is going to fire me otherwise?" Seems reasonable? Further studies showed that any reason was as good as imminent job loss. "May I make some copies now because I'd like to make some copies now?" did equally well — and was significantly more effective than — "May I make some copies now?"

Why are we such suckers? The answer might lie in the selfish nature of our altruism. Recall that favors are granted for mutual advantage. Any statement of need indicates a higher likelihood that the benefit to the receiver is high and the eventual return will be greater.

Another devious experiment reveals a calculating aspect of our favor-granting decision process. Students at the Princeton Theological Seminary were asked a series of questions about their personality and level of religious commitment. They were then sent across a college campus.

Along the way, they met a person who was slumped over, coughing, groaning, and asking for medical assistance. Did self-proclaimed nice people help more? Absolutely not. Neither did religious commitment help explain who was naughty and who was nice.

There was only one robust predictor of altruism. Half the seminarians were manipulated to think that they were late for an appointment while the others were told that they had plenty of time. Sixty-three percent of those with spare time helped as compared to only ten percent of those in a hurry. When they were short of time, even those who reported "religion as a quest" did not stop to help.

Does this mean we are all bad people? No. It means that our level of goodness and badness depends on the payoffs both to us and to the recipient. In the copier example, an indication of high need elicits altruism, whereas in the cross-campus trips, the low cost of giving, in the form of having spare time, is the prod.

The implications, minor and major, of these selfish origins pervade our lives. Julia, a friend of Jay's, recently wanted to thank him for some help. She sees him every day and they exchange numerous e-mails as well. She could have included a verbal or written thanks. Instead, Julia wrote a note and walked past Jay's office to post the letter through the old-fashioned U.S. mail. Knowing that people are sensitive to costs and benefits, Julia knew her thanks would be more appreciated if she expended some effort.

Perhaps the most practical, if trivial, finding is that we should smile and wave at others in traffic. Jay's wife, Lisa, is a master at getting people to allow her to merge into their lane. She makes eye contact, smiles, and asks for permission using facial expressions. Other drivers seem happily magnanimous as they let her pull in ahead of them. Why? The eye contact with Lisa and the facial expressions feel like the beginning of a relationship, thus stimulating our favor-granting instincts.

Since favors are granted, at their core, in the expectation of being repaid, Lisa's acknowledgment of a debt lubricates the roadways. In the enormous city of Los Angeles, the odds that she'll actually have the opportunity to return the favor to this particular person are zero, but this doesn't diminish the effect.

❦

We are living with outdated social instincts in the modern world. One interesting twist to our outdated social instincts is a systematic overestimate of our importance. Pundits lament disappointing voter participation, commenting on the record low turnout of 49% in the 1996 U.S. presidential election. But the real puzzle is why someone would devote an hour to vote in a presidential election at all.

It is impossible for any single vote to change a national election, so the selfish, rational thing to do is stay home. This is all true in 2000, but it most definitely was not true for our ancestors, who lived in small bands and were each as powerful as a modern senator. One voice among a handful could be heard.

The genes in our brain still think that it is completely reasonable to spend some effort swaying the outcome toward their own advantage. They also lead us to care more than we ought to about our multimillion-person society. Because people lived in small groups for so long, our genes have built us with instincts appropriate for a world with a small number of people who interact frequently. In addition to the voting booth, these "small world" instincts affect behavior on the highway.

Aggressive and violent driving is so common that in a recent survey Americans rated road rage as a bigger threat than drunk driving. Numerous websites exist where people can post their stories. One man boasted that after an older woman cut him off in traffic, he followed "the old bag home and bashed her mail-box down."

Why do we get so worked up if the guy in the next car gets ahead of us? Is it really crucial to get to work eight seconds earlier? The answer may again lie in our instincts assuming that we live in a small world.

Honor and reputation are crucially important when we interact repeatedly with the same people. Early in Nelson Mandela's prison term, for example, the guards tried to get all the inmates to run. Mandela exhorted his colleagues saying, "Don't succumb to these threats. Just walk at your normal, steady pace." He explained later that he knew if he had run that day, he'd have been running every day.

When we will be seeing the same people over and over, establishing a reputation can be worth a lot. When we take this attitude onto the highway, however, we are taking risks to punish people whom we will most likely never see again.

Even today, a significant number of fights take place in bars and a surprising number escalate over trivial matters of honor. In fact, over 20,000 Americans will die in homicides this year, and tens of thousands more will be injured in stabbings or gunfights that could have ended in death. A significant percentage of these incidents will occur between strangers who would never have seen each other again had they just walked away.

We vote even when we can't alter the outcome because we expect our vote to register with ancestral impact. Similarly, we are too eager to defend our honor even when honor ought to be forgotten. When we are getting all worked up over some insult, we need to count the number of future interactions with the miscreant. If the answer is zero, we are better off ignoring the slight.

The movie *Full Metal Jacket* chronicles U.S. Marines undergoing initial military training, then follows them to Vietnam. During boot camp, the drill inspector performs a nightly review of his troops and their quarters. One evening he is incensed to find that Private Pyle has not locked up his gear. In a fury, he scatters Pyle's possessions and says, "If there's one thing in this world that I hate, it's an unlocked footlocker! . . . if it weren't for dickheads like you, there wouldn't be any thievery."

An example of blaming the victim? Definitely. But also a commentary on human nature. Because people might be tempted by our possessions, we can help others quell the criminal within by providing an environment that favors honesty.

Similarly, recognizing the selfish nature of friendship can, perhaps paradoxically, strengthen our relationships. Each of us has a unique set of genes. Their ruthless self-interest leads us into conflict with strangers, friends, and even our families. Understanding that genetic self-interest underlies both conflict and cooperation, we can construct situations to induce cooperation.

Friend and foe are fluid categories. Because cooperation is driven by mutual interest we shouldn't be completely averse to seeking and cultivating opportunities with our antagonists. We should learn to be nicer to our rivals because they may be friends or spouses next week. Similarly, it pays to be more guarded with allies. Weaknesses we reveal may be used against us in the (near) future. Finally, we should be nicer to ourselves, our only permanent ally.

Conclusion Surviving desire

Herring gulls lay their eggs on the ground in shallow nests. Though they care dearly for their young, they don't build particularly good nests. The problem with these ramshackle homes is that precious eggs have a tendency to roll out of the nest, where they may be eaten or perish in the cold. Accordingly, the first thing a gull does when returning home is look for stray eggs and push them back into the nest.

Scientists manipulated the number and size of eggs outside the nest and discovered a simple behavioral pattern. The gulls roll the eggs back one at a time and always in the order of biggest to smallest. The scientists continued to tweak this system by making artificial eggs that looked like the natural eggs, but were larger. When they were placed near the nest, the conscientious parents continued to retrieve all the eggs, artificial and natural, from biggest to smallest.

In a twist that would make Pamela Anderson Lee proud, the scientists made enormous fake eggs. The gulls, it seems, have no upper limit to their "bigger is better" rule. Even when the artificial egg was much larger than adult birds themselves, they still tried to save the biggest first. Unable to move an almost football-size artificial egg, the parent nevertheless tried relentlessly and persisted even as its real babies died nearby in untended eggs.

Why would evolution design such a stupid bird or allow it to survive? In fact, the gull's instincts function beautifully in its natural environment. "Bigger is better" works flawlessly because in the real world a gull is never going to encounter a gigantic fake egg, and bigger eggs produce healthier offspring. The problem arises only when the birds are placed in an unusual environment run by meddling scientists.

Like the herring gulls, our instincts worked well in our natural environment, but get us in trouble in an industrialized world. A prime example is our love of eating. Ancestral humans were always hungry, having no reliable food source and no refrigerator or storage system. Their survival rule was simple: eat as much as possible. When we follow this rule in our rich, modern world, many of us become overweight and unhealthy.

Our outdated genes frequently get us into trouble. As with our hearty appetites, many of our problems are simply wanting (and getting) too much of a good thing. What is useful in small quantities often becomes destructive in excess, so instinctual desires in a new environment lead us straight to a problem. In other cases, the source of our trouble is less direct. Consider how satisfying the biological needs of the !Kung San got them in trouble.

Until just a few years ago, the San lived as our ancestors did, hunting wild animals and gathering plants. Some of the first Westerners who contacted them in the 1960s asked what they wanted. For the San the answer was obvious: water. They live in a desert and are perpetually searching for water. Even a slow drip from one of our faucets would provide enough for a small band of people.

Lo and behold, there is plenty of underground water in the Kalahari. In 1962 the Westerners drilled five boreholes in the area known as !Koi!kom, thereby providing a stable water supply. Unfortunately, these San simply traded one nightmare for several others. Normally, they are nomadic, moving from place to place as animal migration or plant seasons dictate. With their boreholes, the San unpacked and settled down nearby. Soon they had depleted all the animals and plants within practical walking distance.

Furthermore, the San had never needed to develop any sanitation methods, leaving their garbage and bodily waste just outside their huts and fireplaces and moving on before debris could build up. Mother Nature took care of recycling. Wedded to their water and unwilling to move, however, they found that their waste piled up and began causing illness. Satisfying the San's water dreams quenched their thirst but made them hungry and sick.

The problems of the !Kung San and the herring gulls illustrate the intricate balance between an animal's instincts and the environment. Today we each face more profound versions of these problems. Our love of possessions, food, and generally easy living has moved us far from our natural setting, creating a plague of troubles in the process.

Our world is changing with dizzying speed. A new computer is outdated by the time it is installed, and a week seems like an eternity in the Internet world. In contrast, evolution is plod-dingly slow, and human genes have not changed very much in thousands of years. Plato would have been puzzled by e-mail, but he enjoyed the same buzz we do from a fine glass of wine. His brain contained exactly the same genetic pleasure buttons that we have.

In fact, our genes are largely unchanged from a time long be-fore Plato. Genetically, we are still cavewomen and cavemen despite our living in ultramodern homes. This mismatch be-tween our genes' natural world and the modern world causes many problems. Drug addiction, obesity, gambling, and bank-ruptcy do not, however, stem simply from innocent discord be-tween ancient and modern worlds. The explanation is more sinister.

People profit from exploiting our outdated instincts. Humans and other primates, for example, love fruits because they are naturally loaded with sugar. Food manufacturers pander to our sweet tooth. While an orange is 10% sugar, some breakfast cere-als have been pumped up to more than 50%. So one of our an-cestors would have let out a whoop of joy to find a naturally-sweet orange, but our children prefer Cap'n Crunch.

Similarly, fast-food pushers did not create our taste for fatty, salty, calorie-laden foods, they simply exploit our existing de-sire by producing a product with exaggerated features. Our taste buds go crazy for a meal that has more, more, and finally even more of the ingredients that kept our ancestors going. The list of profit-making, instinct-exploiting products is long. Por-nography takes advantage of our sexual interests. TV soap op-eras satisfy our taste for social information. And so on.

"Greed is good. Greed captures the essence of the evolutionary spirit and has marked the upward surge of mankind." Or so says Gordon Gekko in *Wall Street*. But he is wrong.

Greed is neither good nor bad, it simply seeks profit. This motive gives rise equally to life-saving vaccines and exploitative loans that charge interest rates north of 100%. What all products — both helpful and destructive — have in common is that they tap into our instinctual desires. It is precisely these desires that so frequently get us into trouble.

In a somewhat cruel test of the human ability to control ourselves, psychologists placed single marshmallows in front of four-year-old children. One of the scientists would then say, "I am going to leave and return in fifteen minutes. You may eat your marshmallow now, but if you wait until I return, you can eat two marshmallows." Hidden observers recorded the kids' initial struggle to resist their urges to eat. Nonetheless, most succumbed and ate the lone marshmallow.

The elegance of this experiment is that the same children were visited more than a decade later. Those who had shown willpower in the marshmallow experiment were more successful than their less disciplined classmates. They were rated as better able to concentrate, more adept at coping with stress, and actually scored significantly higher on the SAT.

All of us face daily marshmallow battles, and bountiful rewards accrue to those of us who can best control our passions. On the road to our dreams we must drive past many alluring detours.

This reminds us of the central theme of *Mean Genes.* The enemy that makes us love fatty foods, look with desire at our neighbor's spouse, and travel hours to risk our paychecks in casinos lies in our own genetic desires.

As tough as our self-control battles are, we at least have a fighting chance. Most animals, even intelligent chimpanzees, have no ability whatsoever to override their passions. Here's a trick researchers teach chimps. In one hand, the researchers hold something the chimps want. They will only give the item to the chimp, however, if it points to the researcher's other hand. Chimps quickly learn this little game and point to the researcher's left hand to get their booty from the right hand, or vice versa.

This ability to learn disappears, however, if the desired item is food. When the chimps see food (a juicy banana, for example), they go straight for it, forgetting the game completely. Even after dozens of failures they keep pointing, with growing frustration, at the hand with the food they desire and don't get. Chimps simply cannot use their intelligence to override their passion for food.

As difficult as willpower is for humans, our capacity for self-control sets us apart from the rest of the animal kingdom. So, in addition to genes that get us in trouble, we have genes for free will and self-discipline. It is within our very genes that we find the tools to fight our animalist urges and take control of our lives.

There are multiple routes to reining in our passions. Let's call one Arnold in honor of the pure discipline shown by Arnold

Schwarzenegger. As a teenager, he set out to be the world's best bodybuilder. Through iron willpower he ruled over the muscle world and parlayed that success into a movie career and more.

Many self-help strategies are variants of the Arnold approach. They ask us simply to get tougher, to live among temptation but to be strong. There is beauty to this approach. We respect discipline. In the James Bond movie *Moonraker,* the villain has a pair of impressively trained Dobermans that patiently ignore juicy steaks just under their noses until given permission. The scene is striking because the ability to resist urges so completely is rare.

But beyond its sheer difficulty, the Arnold approach has the drawback of requiring continual vigilance. If, after a full day of resisting gnawing hunger, we break down and eat a chocolate bar that contains sixty grams of fat and five hundred calories, we go to sleep feeling bad and resolving to be tougher tomorrow. Twenty-three hours and fifty-nine minutes of discipline can be undone in a moment of weakness.

Some temptations are better avoided than resisted. To be sure, we can all benefit from pure mental toughness. Those of us, however, who will eat the marshmallows and scarf down the juicy steaks can benefit from additional tools.

Recall from our discussion of drugs that the intense pleasure we feel during an orgasm or a hit of crack cocaine is molecules of dopamine tickling our brain's "do-it-again" center. Imagine a product that could reliably produce this buzz without side effects. No destructive urges, no HIV-infected needles, just a short-term hit of dopamine free of consequences. We have such a product. It's called a roller-coaster.

Our ancestors got their risk-generated dopamine the old-fashioned way: they took risks. With a bit of ingenuity, we have invented products that create an illusion of risk. Horror movies, bungee jumps, and action video games all give us a risky thrill yet are no more dangerous than taking a nap on the couch. No wonder we like them so much. For other problematic passions, we haven't yet produced such effective products. The potential lies largely untapped. Let's consider a couple of promising approaches, though.

Food substitutes seek to allow us to have our dietary cake and eat it, too. Nutrasweet is one prominent chemical (and there are dozens more in development) designed to fool our taste buds. It promises us "all of the pleasure, none of the costs." There's no barrier to eventually having meals that taste like a four-thousand-calorie steak- and potato-fest but are as healthy as broccoli with brown rice. Similarly, both nicotine gum and methadone try to satisfy our drug cravings while minimizing the associated ill effects.

Innovation can thus help tame our primal instincts. We can make products that stimulate our ancestral instincts but have whatever effect we choose. Junk food becomes health food. Danger on a roller-coaster is a safe thrill. Warfare is conducted with soccer balls and hockey pucks. Cigarettes are replaced by nicotine patches.

A second route to self-control is illustrated in the movie *There's Something About Mary*. In it, Ben Stiller has a history of bad dates in which he gets so excited around attractive women that he scares them away. When he secures a date with attractive Cameron Diaz, he worries that overexcitement will ruin his big chance. A friend advises him to remove the "baby batter" from his brain. Stiller masturbates just before the date and, in his cooled-down state, gets the girl.

Stiller's pre-emptive strike demonstrates another *Mean Genes* tool in our self-control belt. Before we get into a situation where our passions are likely to lead us astray, we can take steps to alter those passions. For example, we temper our food passions by eating something healthful before going to a party or grocery store. Or we consume Antabuse to make drinking alcohol unpleasant.

⌣

Self-control battles have plagued all humans, and many of our oldest legends revolve around this theme. One of the most enduring adventure stories is *The Odyssey,* Homer's description of Odysseus returning home to Greece after the sacking of Troy. Among the perils he faced were sea nymphs, the Sirens, who sang so beautifully that mariners were compelled to approach and inevitably crashed their ships on the surrounding rocks.

Odysseus lashed himself to his ship's mast, plugged the ears of his crew with wax, and gave them strict instructions to ignore his facial expressions. In most situations, having more freedom and power is a good thing. While Odysseus was hearing the Sirens, he could neither move nor order his crew to take him too dangerously close to their lair. It was precisely his planned powerlessness that saved him from destruction. He became the first person to hear the beauty of the Sirens' singing without perishing.

Odysseus anticipated his weakness and took steps to prevent his predictable passions from wreaking destructive ends. The drama of the Sirens is played on smaller stages when we decide to buy lower-fat foods to stock our pantries or restrict lunch dates with an attractive, flirtatious co-worker to public venues.

If we had Arnold-like muscles of discipline, we could choose not to eat even in the midst of chocolate bars. Alternatively, we can outsmart our passions by making sure that only rice cakes are available when the urge to binge strikes. The Chinese philosopher Sun Tzu said, "Those skilled in war bring the enemy to the field of battle and are not brought there." Similarly, our self-control struggles are frequently decided by the terrain; we should pick a setting in which we will win.

Odysseus also teaches us to enjoy life. He could have avoided death simply by putting wax in his ears, as his crew did. But he sought to experience the rich intensity of the Sirens' song and avoid the downside. Our desires create difficulties, but without pleasure, what's the point?

We should enjoy our animal passions and even indulge them but prevent them from controlling us. The key to a satisfying life is finding a middle ground that combines free-flowing pleasure, iron willpower, and the crafty manipulation of ourselves and our situations.

Our temptations are powerful and persistent, but we are not destined to succumb. Ancient and selfish, our mean genes influence us every day in almost every way. But because we can predict their influence, self-knowledge plus discipline can provide a winning strategy in the battle to lead satisfying and moral lives.

Acknowledgments

Mean Genes draws on the research of thousands of scientists. We wish to thank a handful whose work has inspired us and influenced our thinking: David Buss, Napoleon Chagnon, Leda Cosmides, Martin Daly, Nicholas Davies, Richard Dawkins, Irv DeVore, Jared Diamond, Peter Ellison, Helen Fisher, Robert Frank, Jane Goodall, Kim Hill, Sarah Blaffer Hrdy, A. Magdalena Hurtado, Daniel Kahneman, Melvin Konner, John Krebs, Randy Nesse, Steve Pinker, John Tooby, Robert Trivers, E. O. Wilson, Margo Wilson, and Richard Wrangham.

We are indebted to a set of mentors who shared both their knowledge and their ways of thinking. Steve Austad, Richard Lewontin, and Michael Rose played central roles in Jay's intellectual career, as Adam Brandenburger and Vernon Smith did for Terry. We both were fortunate to learn directly from Irv DeVore, Peter Ellison, and Marc Hauser.

E. O. Wilson deserves particular thanks. His seminal writings played a central role in each of our careers. Furthermore, he has been a kind yet demanding mentor to both of us, and he has shepherded *Mean Genes* along at important stages.

Many friends donated time reading drafts and providing crucial feedback. John Fetterman has read and critiqued every chapter (some of them several times). Alicia Moretti has similarly been involved at each stage, bringing her linguistic love to bear. Julia Vallone read numerous chapters, often on short notice, and kept us going with her intelligence and humor.

Throughout every stage of the book's development, Lisa Phelan has been a steadfast supporter and a fountain of writing and publishing advice. Her thoughtful input and unwillingness to be pessimistic for even an instant helped us to see what a gem each day could be.

In addition, others who made substantial contributions include: Glenn Adelson, Kim Alley, Nicole Belle Isle, Ben Berger, Jeff Bodenstab, Riley Bove, Thomas & Marie Burnham, Katie Cahill, Nancy DeVore, Sue Flewelling, Judith Flynn, Kate Gawronski, Joan Greco, Brian Hare, Carole Hooven, Matt Krepps, Chris Matheson, Fatima Melo, Matthew McIntyre, Michelle McNamara, Boaz Moselle, Harold Owens, Kathleen & John Phelan, Kevin Phelan, Patrick Phelan, Michelle Richmond, Ilene Rosin, Emma Schiffman, Dani Schindler, Barbara Li Smith, Bill U'Ren, Kathleen Valley, and Mike Walfish.

We thank our agents John Brockman and Katinka Matson for being forceful advocates on our behalf. As neophytes in the publishing world, we also sought and received valuable advice from Holley Bishop, Steve Pinker, and Barbara Rifkind.

Amanda Cook, our editor at Perseus, shares our vision for the book. In a world in which few editors really edit, Amanda has pushed, prodded, and proposed. She made major contributions to the book and has earned our respect and gratitude. Also at Perseus, David Goehring has been a potent and fun champion for the project.

Elizabeth Nowlis, Laurie Puhn, and Julia Vallone dug through library stacks and cyberworlds to help ensure that every fact was checked and every statement is true.

Thank you all.
Terry & Jay

Notes The data in *Mean Genes* have been assiduously researched and documented. If you want to learn more, we encourage you to visit the *Mean Genes* website, www.meangenes.org, where you will find a set of notes.

Index